CURRICULUM GUIDE

GEOGRAPHY

One of a series of six rivers-based units written by teachers participating in the Rivers Curriculum Project funded by the National Science Foundation

Dr. Robert Williams, Project Director
Southern Illinois University at Edwardsville (SIUE)

Cynthia Bidlack, Project Coordinator, SIUE

Lead Author:

Bob Ashley, Red Bud High School

Contributing Authors:

Jack Herschbach, Chester High School
Dennis Paul, Eisenhower Decatur High School
Larry Underwood, Brussels High School
Cynthia Bidlack, Jersey Community High School

Dale Seymour Publications®

Managing Editor: Catherine Anderson
Project Editor: Christine Freeman
Production Coordinators: Barbara Atmore and Joe Conte
Design Manager: Jeff Kelly
Text and Cover Design: Christy Butterfield
Cover Photograph: Nicholas Pavloff
Illustrations: Cyndie C. H. Wooley
Technical Art: Jonathan Clark

This book is published by Dale Seymour Publications®,
an imprint of Addison Wesley Longman, Inc.

Printed in the United States of America

Order Number DS30616

ISBN 0-201-49368-3

Daye Seymour Publications®
P.O. Box 5026
White Plains, NY 10602-5026

This book is printed on recycled paper.

1 2 3 4 5 6 7 8 9 10-ML-00 99 98 97 96

Acknowledgments

Rivers Project Curriculum

We cannot list all the teachers, scholars, and friends to whom we are indebted for the development of the Rivers Project curriculum units, but we want to mention some of these very important people.

To start, without the help and guidance of Mark Mitchell and Bill Stapp, GREEN Project at the University of Michigan, Ann Arbor, in the very early stages of the Rivers Project, we might not be where we are today. The same thank-you goes to Tanner Girard, Illinois Pollution Control Board Judge and former professor at Principia College in Elsah, Illinois, who has been helpful in so many ways. We owe so much to Don Humphreys and Ivo Lindauer, our National Science Foundation Program directors, who have believed in and supported the curriculum development since its inception and throughout the long and tedious job of writing, piloting, rewriting, and field testing. We offer special thanks to the Illinois State Board of Education and the Illinois Higher Board of Education for beginning the project and for offering continuing support for Dwight D. Eisenhower teacher training.

No way exists for us to thank sufficiently each of the curriculum unit writers for the hours and hours of personal time devoted to their specific unit. Not only have they been writers but also trainers for new teachers entering the program, during both the school year and the summer training sessions.

We want to thank all the curriculum unit users for revision suggestions given to the writers. We are especially grateful to them for taking time away from their traditional classroom setting to give time to their students out on the river for hands-on activities.

To the university professors from across the country who read the units in their area and provided professional opinions and suggestions for improvement, we are sincerely thankful.

Finally, to Pat, Jack, Bill, Michele, and all the many university student workers and graduate assistants who have come and gone over the course of the last four years, what would we have done without you?

Dr. Robert Williams, Project Director
Cindy Bidlack, Project Coordinator

Cynthia Lee (Cindy) Bidlack died of cancer shortly after her 46th birthday. She gave so much to this project. For all her efforts, we can only say thanks. We remember her often and miss her much.

Rivers Geography

The geography curriculum team gratefully acknowledges the contributions of the educators whose advice and assistance have been of great importance to this project. In particular, Sarah Bednarz (College Station, Texas) and Gary Miller (Virginia Beach, Virginia) read early drafts and provided detailed, insightful suggestions. Gerald Speed (River Ridge High School, Elizabeth, Illinois), one of numerous teachers to pilot the curriculum, wrote extensive notes after testing the lessons in his classes. Fred Walk (Normal Community High School, Illinois), Dr. Norm Bettis (Illinois State University), and James Marran (then of New Trier High School, Winnetka, Illinois) contributed clarifications at times of great need. The influence of Dr. Fred Lampe (Southern Illinois University at Edwardsville), a creative geography educator and pioneer in the promotion of geography's five themes, was present throughout the project.

CONTENTS

RIVERS CURRICULUM PROJECT

When many of us think of rivers, we picture the fun times we have around them—boating, swimming, frolicking, watching wildlife, just enjoying being where land meets water. We may also appreciate the benefits that rivers give to us, such as drinking water, plentiful food sources, recreation, electric power, and an efficient means of transportation. While tallying up all these good things about rivers, however, we may grow concerned about threats to the health of our rivers.

Rivers and streams confront many forms of pollution—industrial waste, acid rain, sewage spills, and thermal pollution. Fortunately, scientists, environmentalists, and the public have instigated regulatory and technical changes that seem to be reducing some of these risks. What kinds of river pollution, and effects of pollution, need further attention? What other threats do rivers and streams face?

Scientists increasingly tell us that the main threat to America's rivers today comes not from pollution, but the physical and biological transformation of rivers and their watersheds. As rivers are altered to provide water transportation, generate power, reduce flood hazards, and provide water for farms, cities and industries, their physical, chemical, and biological processes are damaged or destroyed. The loss of riverside and aquatic habitat has led to the decline or extinction of more than one-third of North America's fish species and an even higher proportion of its native mussel species.

Healthy river systems are incredibly dynamic. As nutrients, sediments and organisms are transported downstream, water and organic materials are constantly added to the mix. Most of these materials come from the surrounding terrestrial system, with the land-water boundary, known as the "riparian zone," acting as a critical valve or filter that regulates the exchange. Riparian zones and their associated wetlands also act as natural sponges, absorbing and filtering polluted floodwaters over time. Where the banks of streams are cleared, straightened, and replaced with rocks or concrete to reduce flooding, the ability of associated wetlands and floodplains to control and filter runoff, provide habitat, and add nutrients is lost.

When rivers flood, they alter the shape of the stream, scouring new channels, inundating riverside land, depositing sediments, and building new banks and beaches. These functions are as important to healthy river ecosystems as natural fires are to healthy prairies and forests. For many fish species, this flood "pulse," called the "natural hydrograph" by scientists, not only triggers spawning and migration but also allows fish to reach seasonally inundated floodplain nurseries and spawning habitats.

Scientists have made significant progress in understanding how changing natural hydrologic cycles has contributed to the destruction of aquatic ecosystems. Numerous local communities have taken the lead in adopting cost-effective storm water, floodplain, and water-supply management programs that utilize natural hydrologic processes. Sometimes, however, federal and state agencies want to alter the hydrology and other physical characteristics of rivers and their watersheds. The debate on such issues ranges down many river corridors.

In order for the youth of our country to become informed participants in the political process, they must have a solid background in environmental issues. The school systems must, therefore, educate students about the nature of the environment. Through the study of rivers, not only does a concern for water become important to the students, but other issues begin to gain this same importance. And for our country, and our world, this can only be one giant step forward.

The Rivers Project curriculum is the end product of four years of environmental commitment by hundreds of high-school teachers in the United States and Canada. Because of the growing importance of environmental issues, teachers through whose towns a river or stream flows have sought to integrate water and river studies into their traditional content courses.

USING THE RIVERS CURRICULUM

When students visit a river or stream and become actively involved in observing, measuring, testing, and writing about that waterway, they quickly develop a sense of ownership toward that river or stream. They also tend to develop a broader understanding of the value of the academic discipline that has brought them to the water's edge. Toward these goals, the Rivers Project units were developed by teachers for use in science (chemistry, biology, earth science), geography, language arts, and mathematics classes. The Rivers Project curriculum prepares students to perform field investigations, with the primary laboratory being a local river or stream. Some activities are to be used in the classroom, focusing on preparation for the field experience, which may be one long trip or several shorter ones. Extensions of the curriculum units encourage teachers to make use of other kinds of field trips, to invite guest speakers to the school, and to make contact with local, state, and federal agencies for outside resources.

In a number of states, governmental agencies are using in their monitoring processes data collected by Rivers Project students. Such contact with state agencies is a vital component of the recommended scientific activities stated in the curricula for these units. As students experience hands-on learning activities that result in river data of real scientific or cultural value, affecting real-world issues they care about, they are motivated to learn even more.

The Rivers Project curriculum is not intended to be used as a textbook, but as a set of supplementary materials designed to enhance your existing program or to establish a basis for river study in your school. The materials involve students in a natural environment—the river—through a series of hands-on activities conducted through field-based study.

When using these units, let your imagination and your creativity run wild. The materials can prove a valuable addition to your traditional teaching. *Remember, what your students feel and touch as part of their river experiences will stay with them for the rest of their lives. Make it memorable!*

Rivers Project Curriculum Units

The river is the common strand weaving the units together into the interdisciplinary curricula you are about to use. This brief synopsis and the underlying connection of each curriculum unit to the other will aid in understanding how the Rivers Project units may be taught. You may use these units independently of each other, or you may combine them in an interdisciplinary approach, especially a team-teaching venture.

Rivers Chemistry defines water quality and guides students in basic data collection. Water-quality kits, which are readily available and easy to use, make conducting the tests a relatively uncomplicated task.

Rivers Biology focuses on stream-monitoring programs and the study of benthic macroinvertebrates. Living organisms in a river, stream, or lake are easily captured or documented. Their existence and numbers provide data for comparison with those of the chemical unit. Biological diversity for a water environment may change as the water quality improves or decreases.

Rivers Earth Science evaluates the physical features of a river system that provide clues to understanding the historical development within a local area. Students better understand the impact of the river drainage system on water quality when their study also factors in soil, slope, and flow. As scholars in the twenty-first century study the effects of agriculture, development, and transportation on the river and water resources, they will view the geology of an area with increasing attentiveness.

Rivers Geography enables students to develop a sense of the environmental impact of people occupying and organizing themselves along rivers. Study of the geography of the river as it relates to location, place, movement, region, and human-environment interactions along its banks gives form and reason to human migration and development. The river becomes a lab for an ever-changing society.

Rivers Language Arts focuses on the skills students will use as they investigate and write about their study of the river. Lessons include technical writing for scientific reporting, interviewing, research techniques for exploring local history, political letter writing, poetry, and other forms of creative writing. Samples of, and references to, exemplary fiction and nonfiction written about rivers are included. Teachers of every discipline can use this unit to give voices to the discoveries and ideas garnered by their students through river study.

Rivers Mathematics provides real-life application of mathematical processes and skills, using data gathered during field studies and from reference sources. Topics range from measurement and working with percentages to standard deviations. This material specifically teaches the mathematical skills required for the science units in the Rivers Project curriculum.

Scheduling and Team Teaching Options

Each science unit, and the geography unit, can be used as a freestanding unit involving about one month of consistent focus. Alternatively, any of these units can be integrated into the regular activities of the class throughout a longer period. The language arts and mathematics units can be used in language arts or mathematics classes to support and expand river activities being done in the science classes. Science and environmental education teachers who wish to broaden student experience in other areas, or to add an interdisciplinary dimension to their curriculum, can also use these units as support material.

Because each unit can stand alone, a few units do have some duplication of topics. For example, because of its importance, analysis of fecal coliform is included in both the chemistry and the biology units. If teachers in both disciplines are using the Rivers Project curriculum, such testing may be performed in just the biology class, with the results shared with the chemistry class.

A single teacher in a school may use a Rivers Project unit, or several teachers may work as a multidisciplinary team. In many schools, what began as a single unit has grown over the course of several years into a schoolwide project as more teachers and students have become involved and the school has acquired more materials and equipment for river study.

Using Your River or Stream in This Curriculum

Teachers and students can use the Rivers Project units on any river or stream anywhere in the world. The constants are the water tests and educational studies that student perform on the river. Each river or stream presents a unique set of parameters for collecting and studying. Investigators approach bigger rivers very differently from smaller streams or shallow rivers. A cold mountain river displays vastly different flora and fauna from the slow, sluggish coastal river. Factors such as population density, natural and modified drainage patterns, and climate vary not only from river to river but from one point along a river to another. The study of each river and its corresponding watershed, therefore, has special attributes that cannot be dealt with in generic materials such as these.

You and your students will want to build a library of handbooks, field guides, and other support material appropriate to your specific area. The best sources for this information are local agencies that monitor and study the waterway. Soil and water conservation district offices, state and local conservation departments, environmental protection or water agencies, and local universities are good places to begin. Most state agencies have a number of relevant free publications. While building the library, locate a person who can advise the class in the field. Many schools have found locally based state employees who are willing to provide content support, sometimes equipment, and even to join the class on water-monitoring days. Once your class has a mentor, river study often becomes easier and more meaningful for you and your students. More than likely, the data collected by the students will also be important to the mentor.

ABOUT THE RIVERS PROJECT

The Rivers Project curriculum in these six units started with teachers who loved teaching and loved the rivers in their communities. They banded together to spread their enthusiasm, knowledge, and learning tools among students and other teachers.

History of the Rivers Project

The Rivers Project began as the Illinois Rivers Project in February 1990 as a pilot program involving eight high schools along the Mississippi and lower Illinois Rivers. With scientific literacy as the ultimate goal of the Project, students collected and analyzed water samples from test sites along both rivers. The study of the rivers was extended to include historical, social, and economic implications of the state of the rivers, thus involving students from classes across the curricular areas of science, social studies, and English.

SOILED NET, a telecommunication network linking the participating schools with each other and with Project headquarters, provided a technological framework for many of the Project's activities.

In December 1991, the Illinois Rivers Project received a grant from the National Science Foundation to develop a formal Rivers Project curriculum. The resultant units, in chemistry, biology, earth science, geography, language arts, and mathematics are applicable to any river in the world. To further the Rivers Project, Southern Illinois University at Edwardsville hosts a training session each summer. At these programs, teachers who use Rivers Project materials serve as mentors for participating teachers in classroom and field-study sessions.

Goals of the Rivers Project Curriculum

The three principle goals of the Rivers Project curriculum are:

1. to increase students' knowledge and understanding of important issues and concepts related to the river.
2. to prepare students with the necessary skills to properly investigate and report relevant information regarding the river.
3. to inspire students to take action to resolve problems that contribute to the overall deterioration of the natural beauty and functions of the river.

Project Funding

The Rivers Project began with funding from the Illinois State Board of Education through a Scientific Literacy Grant. Subsequent funding for Illinois teacher training was obtained from the Illinois Board of Higher Education, Dwight D. Eisenhower Title II funds. A grant from the U.S. Fish and Wildlife Service allowed Midwestern schools in Iowa, Minnesota, and Wisconsin to participate. The National Science Foundation, under its Materials Development Program, provided funds for the final development and testing of the Rivers Project curriculum units. Along the way, many others in the academic and business community, as well as numerous local, state, and federal agencies provided support.

TELECOMMUNICATIONS

A key component of the Rivers Project is the sharing of water-quality data and student writings among schools in the project. Though schools can communicate by mail, use of the computer networks associated with the Rivers Project provides for faster, more responsive, more directly accessible exchange. Students using these networks can not only contribute their data but also quickly access the data, writings, and queries of students at other Rivers Project schools, most notably students exploring the same river watershed.

The Rivers Project utilizes two telecommunications systems, e-mail and the World Wide Web. Either requires a computer and a modem. Because telecommunications is evolving rapidly, the telecommunications technology, data-handling, and data availability for the Rivers Project will undoubtedly also change. For the most current information on how to contact the Rivers Project, telephone Southern Illinois University at Edwardsville (SIUE) at 618-692-3788 or 618-692-3065.

E-Mail

Students using a computer that has a user account can send writings to the Rivers Project via e-mail using the address "rivers@siue.edu." Other students may read these on the World Wide Web. The Rivers Project also periodically publishes collections of student writings it has received.

World Wide Web

The Rivers Project is also available via the World Wide Web. This requires web browser software and access to the Internet. The URL (Uniform Resource Locator) is "http://www.siue.edu/OSME/river." This home page contains a searchable database of all water-quality data sent in to date and also a form that can be filled out online to report water-quality results. Selected student writings are also available via this home page. The Rivers Project home page contains a link to the e-mail address.

The Rivers Geography Unit

Geography is for everyone. It enables us to better appreciate Earth, the blue planet, our home. Geography helps us see the interrelationships between human and physical systems, and it stimulates our imagination and curiosity about the world's physical and human diversity. It provides a basis for making decisions about local and global issues.

The chief goal of *Rivers Geography* is to help students become geographically informed. The geography unit will help students understand the relationships among people, places, and environments and the interactions that occur on local, regional, and global scales, especially if the unit is used as part of a broader geography curriculum emphasizing the same goals and applications.

The acquisition of geographic knowledge, skills, and perspectives is a lifelong process. The geographically informed person must continually learn about and adjust to changes in the world, people, and technologies. Through this curriculum, students can better understand and adjust to change.

COMPONENTS OF RIVERS GEOGRAPHY

Rivers Geography focuses on generating and channeling student involvement into a geographic inquiry about a local river or stream. Each of the seven lessons in this unit helps students investigate a geographic question that relates not only to their specific river or stream but also to larger geographic concepts. As a whole, these lessons teach and apply the major themes, standards, perspectives, and skills of geographic study.

To achieve this, each lesson has several components. At the beginning of each lesson, a Teacher Notes segment provides information on how to prepare for and carry out each lesson. Teacher Presentations provide detailed directions for teacher demonstrations; occasional Teacher Guides provide step-by-step procedures and answers for use in parallel with specific student activities. Student Information sheets provide students with information on geographic concepts and procedures, together with opportunities for students to practice and demonstrate this knowledge. Student Activity sheets ask students to perform participatory tasks, such as mapmaking, group discussion, field site observations, analysis, and report preparation. Students should write their responses to questions and assignments for these sheets on loose-leaf paper, to be turned in or inserted in their notebooks. Assessment activities are also included as appropriate (discussed later in this introduction.)

. Plan to have students do in class the parts of activities dependent on the use of classroom wall maps and atlases. Some concepts may require review and further explanation before students can proceed. Review should follow each completed task.

GEOGRAPHY AND THE RIVER

Rivers are one of the most basic elements of physical geography; the presence and character of each river has influenced patterns and styles of other natural elements, plants and animals, and human settlement and culture. In turn, humans have affected rivers and streams profoundly, both on purpose and inadvertently.

The Geographic Perspective

Throughout the Rivers Project curriculum, students look at the relationships among natural and human forces in order to understand what makes a particular river or stream unique. Studies in water quality, geologic formations, and plant and animal species provide pictures of the physical elements; studies of land uses such as settlements, agriculture, transportation, and mining evidence human systems. Geography then incorporates such physical and human systems into spatial systems, so we understand how the parts influence one another.

Geography allows students to understand how Earth's surface is organized and occupied. The recognition of the "Where?" of a phenomenon is of great importance. Location is, however, only the beginning of geographic study. Questions such as "Why is a phenomenon located where it is?," "What are its characteristics?," and "To what else is it related?" give students the freedom and obligation to observe, analyze, and evaluate whatever relationships exist relative to their objects of study. Investigations in this unit carry students across conventionally recognized boundaries of academic disciplines into consideration of data from both the physical and social sciences.

Geography's Five Themes

Many geography programs today are organized around five fundamental themes, presented by the Joint Committee on Geographic Education in their 1984 publication, *Guidelines for Geographic Education*. These themes are location, place, movement, human-environmental interaction, and region. These themes, which have won overwhelming acceptance by elementary, middle school, and secondary educators, clarify the scope and organization of geography and provide a "handle" for teachers and students alike. By using these fundamental themes as a framework, Rivers Geography helps students focus on the real world and make some sense of a specific portion of it—their local river or stream.

The students receive an overview of all five themes in Lesson 1 of *Rivers Geography*. As shown in the table on pages xiv and xv, each succeeding lesson focuses primarily on one theme, specifically tied in to a water setting, yet the material is designed to facilitate student realization that the five themes are interrelated. To simply know the five themes misses the point. Instead, the teacher should employ the themes as a way to organize and give structure to the process of geographic inquiry. Through an integration of the five themes, geography can provide a holistic view of the world.

The lesson and theme matrix in this table also shows how the use of these five themes helps structure the study or a river or stream, and how the study of the rivers and streams, in turn, sparks questions that help make the purpose and structure of geographic inquiry clear.

Five-Theme Graphic

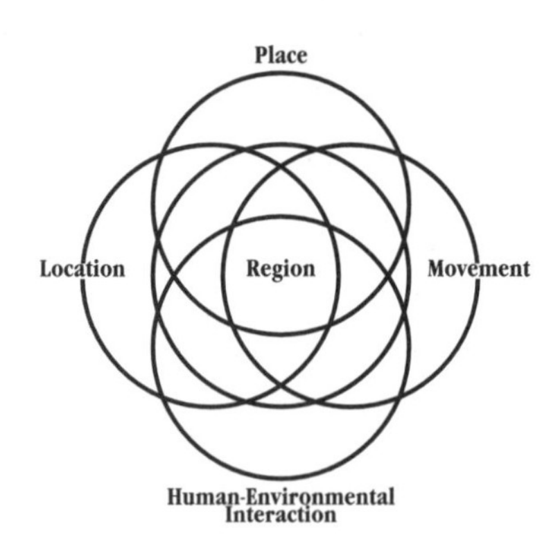

The graphic of the five themes of geography, shown here, illustrates the interconnectivity of geographic subject matter. Though a teacher or student may identify content by a particular theme, separating it from its myriad connections to other phenomena is artificial. Rather than viewing geography as a collection of isolated facts listed by theme, students should learn to see such knowledge as interrelated. A phenomenon takes on a greater significance when its relationships to other themes are described. For instance, geographers may undertake a regional study, investigating the role of several themes in one specific geographic area. Alternatively, geographers may investigate a particular topic, looking to understand in which regions, and in relation to what other phenomena, it occurs.

The five-theme graphic appears on the first page of each lesson, with the focus of that lesson shaded. Use it as a reminder that, even though one or more themes is primary, the lesson pertains to all the other themes as well.

Regardless of the idea with which a lesson begins and regardless of the approach a classroom discussion or project takes, use opportunities to motivate students to discover that, though the region is considered the most complex and central theme to geographic studies, understanding the real world requires using all the themes.

OUTCOMES FOR RIVERS GEOGRAPHY

Through the process of doing *Rivers Geography*, students will make significant progress toward becoming geographically informed persons, as defined in *Geography for Life* by the Geography Education Standards Project. They will have progressed in three interrelated and inseparable areas: subject matter knowledge, perspectives, and skills.

Achieving Geography's National Standards

Practicing geographers and geography educators have identified six essential elements of geographic subject matter, containing eighteen standards to which students should aspire in order to be geographically informed. In *Geography for Life,* the elements and standards are explained in detail and illuminated with examples. *Geography for Life* is a well-written and beautifully illustrated guide essential for the geography teacher at any level.

These challenging standards identify the most important elements of geography, with benchmarks for grades 4, 8, and 12. The standards are activity-oriented, and they describe what a student should be able to do.

Though the standards incorporate the five themes, a teacher cannot simply match up themes and standards. Instead, the teacher (and student) must understand that each theme is essential to understanding and using all the standards. The themes emphasize subject matter; the standards include not only subject matter but skills and perspectives that students should learn. Many teachers of geography are familiar with these standards; for those who are not, they are found in Appendix D.

Gaining Perspective on Geography

Students expand their interpretation of phenomena, and their comprehension of the world, if they learn to see situations from more than one point of view. *Rivers Geography* identifies and supports the multiple perspectives that geography often encompasses.

Geographers most commonly perceive phenomena from two perspectives: spatial and ecological. The spatial perspective promotes an understanding of "Where?" and "Why there?" Understanding the distributions of people and other phenomena, their interrelationships, and the resulting spatial patterns are goals of this perspective.

Geographers use the ecological perspective to understand the relationships of living things—including humans—with one another and with the physical environment. This perspective pays particular attention to human alterations of the physical environment, because human actions have such potential to alter natural ecosystems significantly.

The historical perspective helps geographers increase their understanding of spatial and ecological perspectives. The sequencing of events through the temporal focus helps inquirers understand why a phenomenon exists, why change occurs, and how to make reasonable projections about the future.

Understanding the interactions of people and their societies requires awareness of how people produce, exchange, and consume goods and services. In geographic inquiry, this economic perspective may include employment patterns, population concentrations, standards of living, and resource distribution and development. Awareness of perspectives relevant to a particular geographic inquiry can help channel student thought; as desired, include local examples that underscore these perspectives.

Over the course of the entire unit, the lessons in *Rivers Geography* touch on each theme, standard, element, perspective, and skill. Because these factors are interrelated, do not try to limit teaching focus to just one standard or skill. Attention to a standard, such as "How human actions modify the physical environment," may require students to also understand location, human migration, physical place characteristics, human settlement, culture, and more.

The Teacher Note at the beginning of each *Rivers Geography* lesson lists the theme and perspectives most relevant to that lesson, as well as the geographic standards that receive primary emphasis, and those that receive secondary emphasis, in that lesson. Use these lists simply as a starting point, adding other emphases to suit your teaching preferences and circumstances.

Increasing Geography Skills

In order to become geographically informed, a student must acquire not only knowledge but practical and critical thinking skills for processing that knowledge into a deeper understanding. By completing the activities in *Rivers Geography,* the student should gain significantly increased competency in the skills involved in being geographically informed. The student should be able to:

1. *Ask geographic questions.* Where is it? Why is it there? What is significant about its location? To what else is it related? Identify geographic problems and issues, pose hypotheses, and organize procedures for studying geographic questions.
2. *Acquire geographic information.* Select and gather information from a variety of primary and secondary sources, including maps, periodicals, statistical databases, interviews, personal observations, and photographs.
3. *Organize geographic information.* Design and prepare a variety of maps, charts, graphs, and other graphic displays of information, categorize and

Geography Lessons and Themes

Lesson	Location	Place
1 How to Begin a River Study		How can you use geography's five themes
2 Where Is the River? (Location)	How can you use maps, the geographical grid, compass directions, and distance to show location? Where is the river relative to other phenomena?	What are the locations of selected characteristics of the river environment?
3 What Are the Characteristics of the River? (Place)	What are the locations of selected physical and human features?	What are the physical and human characteristics of the region?
4 How Is Your River or Stream Site Related to Other Places? (Movement)	From what countries or regions did your ancestors migrate? Why did they migrate? Where were you and your classmates born?	What is the ethnic composition of your community? What are the cultural practices that make your community or region unique?
5 How Do Humans Interact With the River? (Human-Environmental Iinteraction)	What are the locations of the simulated communities and regions relative to the river and to one another?	What characteristics can you use to describe and understand the simulated communities and regions? How do they differ from one another?
6 What Can Be Learned From a Field Study? (Field Study)	What are the latitude and longitude of the study site? What is its compass direction and distance from school? Where is the site relative to the watershed and other features?	What physical and human characteristics of the study site region can you observe on a map? What characteristics can you observe during a field study?
7 How Is Your River or Stream Location Unique? (Region)	Where is the study site? What area does it occupy? How do phenomena interact to make the area unique?	What are the physical and human characteristics of the study site? Are they similar or different from those of surrounding areas?

Movement	Human-Environmental Interaction	Region
as a framework to study the river?		
Where in the study region do people live? Where do they work, shop, go to school, and play? How do they get there?	What obstacles or opportunities does the physical environment present? Where are they? How and where have people modified the natural environment?	Where is the study site region? Where is the river relative to the watershed? Of what larger regions are the river, watershed, and study site a part? What and where are they?
What movements of physical and human features can you observe? How and where do they move?	How has the physical environment been changed?	What are the characteristics that make this region unique and distinguishable from others? What gives it character?
How did your ancestors move to your community? What cultural practices did they bring with them?	What changes did the people who settled your community make in the physical environment? How did they use the river?	How have movements of people, ideas, and things changed the character of the river, the region, and the community in which you live? How are they related to other rivers, regions, and communities?
What physical and human movements can you observe in the simulation?	How did the people in the simulation change their physical environments? What were the positive and negative effects?	How did the regional scale influence decisions in the simulation? How do physical and human phenomena interact to create a unique regional personality?
What physical and human movements can you observe or infer?	What human alterations of the physical environment can you observe during the field study? Why? What are the positive and negative effects of change?	What features give the study site region a unique character? How can you describe its personality?
What movements of people, ideas, and things occur within the region and between it and other regions? What are the relationships and connections?	How have humans altered the study site region? What areas have these changes affected?	What is the personality of the study site region?

classify data, summarize, express thoughts in written form, and integrate information.

4. *Analyze geographic information.* Interpret geographic information from a variety of sources, make inferences, draw conclusions, compare, and explain.
5. *Answer geographic questions.* Identify relationships, evaluate solutions to problems, apply geographic models and generalizations, analyze case studies, and state generalizations.

ADVANCE PREPARATION

Before starting *Rivers Geography,* each teacher will need to gather numerous materials, become familiar with topics included in the unit, do preliminary planning for the field study, and assess student readiness for the project.

Student Prerequisites

Rivers Geography can serve as either an introductory unit to geography study or as an enrichment unit; no previous geography study is required. Students should have basic math skills, though algebra is not required. (For specific teaching activities on relevant math skills, see *Rivers Mathematics.*) This unit has been written for high school and middle school students in geography classes. Middle school students may require more instructional time per lesson than older students.

Teacher Preparation

Prior to introducing the *Rivers Geography* unit to your class, you should prepare by doing the following:

1. Read the entire unit.
2. Read *Guidelines for Geographic Education* (National Council for Geographic Education), *7–12 Geography: Themes, Key Ideas, and Learning Opportunities* (Geography Education National Implementation Project), and *Geography for Life: National Geography Standards 1994* (Geography Education Standards Project). (See Appendix C.)
3. Brush up on concepts used in *Rivers Geography.*
4. Coordinate activities and curriculum with teachers considering doing other Rivers Project units.
5. Develop timeline for placing *Rivers Geography* in your school calendar and your yearlong instructional schedule. Times noted in each activity are approximations; allow some flexibility.
6. Discuss the curriculum with your school administration and secure permission for the field study.
7. Select and visit a site suitable for the activities, with easy access and minimal safety hazards. Obtain all necessary permissions from the landowner for students to visit the site.
8. Schedule field-site visit and prepare field-trip permission forms. Secure field-trip transportation. (Begin other steps required before field trip, as listed in Teacher Information 6.1: Field-Study Guidelines.)
9. From the U.S. Geological Survey (USGS) (address in Appendix C, or call 800-USA-MAPS), order:
   - pamphlet *Topographic Map Symbols* (one per two students)
   - topographic map index
   - order form for your state
   - other information about topographic maps,
   - once you know which map to order, order a topographic map of your study area (7.5-minute series) (one per two students)
   - (optional but highly recommended) classroom supplements, "What Do Maps Show?" and "Exploring Maps." Brochures, maps, posters, and classroom activity sheets are included.

   For general earth science information, contact the USGS Public Inquiries Office (address and phone in Appendix C).
10. From the National Geographic Society (contact information in Appendix C), order the poster *Maps, the Landscape, and Fundamental Themes of Geography* (one per two students).
11. As teacher background for Lessons 1 and 2, you may wish to obtain the USGS booklet *Water of the World* by Raymond L. Nace (Water Resources Division address in Appendix C, or phone 800-426-9000).
12. Before beginning and throughout the unit, collect maps, photos, slides, and other material about your field site, community, and region. As well as the USGS, other good sources include your state geological survey, soil and water conservation department, department of conservation, environmental protection agency, and local agencies.
13. Consider ordering *The Language of Maps* by Philip J. Gersmehl. This optional but highly recommended excellent manual introduces students to the language used to express spatial relationships (source information in Appendix C).

14. Order filmstrips and written materials required for the activities in Extending the Lesson you plan to do.
15. Begin collecting sample brochures from travel agencies, government agencies, local tourism-oriented businesses, and local Chambers of Commerce for students to use as models in Lesson 7.
16. Collect all other necessary materials and equipment.
17. Enlist the aid of an expert (such as a conservation officer, planning commission member, Army Corps of Engineers employee, or public employee) who works with river- and stream-related concerns and who is willing to address your students or lead them on the field-site visit.
18. Duplicate all materials to be distributed to students (including Student Information, Student Activity, and Assessment pages from this unit), three-hole punch all sheets, and place in labeled files.
19. Prepare overhead transparencies from masters provided in *Rivers Geography*.
20. Get excited. Your enthusiasm will be contagious.

SUGGESTED SCHEDULE

Rivers Geography is designed primarily for use as a free-standing unit in which a class works through all the lessons and activities in the order presented to form a unit of study. Using this approach, *Rivers Geography* would involve approximately 30 class periods of 40–50 minutes, plus one field-site visit to a river or stream. The length of time required varies depending on the types and amount of assessment activities included. If possible, an additional field-site visit at the beginning of the unit inspires students and increases enthusiasm for geography. To guide student activity on this field trip, consider a stream walk, mentioned in the Teacher Notes for Lesson 6 under Extending the Lesson. (Field-trip information on safety and planning is also found in Lesson 6.) The unit also can be expanded into a semester- or year-long course by pursuing the study of the local river or stream in more depth and including lessons from the other Rivers Project guides.

SAFETY

Before taking students on a field study to a river or stream, provide them with appropriate safety guidelines and make sure they understand and agree to follow them. An example of a student guide on field safety rules is included in Student Information 6.1.

When visiting the river or stream, bring and have students use the appropriate safety equipment listed in Student Activity 6.3.

ASSESSMENT

Assessment is the opportunity for students to demonstrate what they have learned and to exhibit their abilities to apply that knowledge in a meaningful way. The instructor has the responsibility to provide instruments that give students the opportunity to best show their knowledge and skills. *Rivers Geography* provides an assortment of forced-choice and performance-based assessment tools.

Forced-Choice Assessment

Traditional assessment tools assess student knowledge of a subject in terms of how well the student has learned and can reproduce the facts provided in the lessons. In *Rivers Geography,* many of the short-answer questions at the end of Student Information Sheets and Student Activity Sheets are designed to stimulation student acquisition of knowledge and provide information to be used in class discussions. Teachers use these as in-class assignments and as homework. Sample answers for these sheets, which also include short writing assignments, are included in Appendix B. Appendix A contains a forced-choice unit test on *Rivers Geography.*

Performance Assessment

Educators increasingly realize the importance of assessing the extent to which the student can apply the instructional information to problem-solving situations; in other words, teachers want to know whether their students can really use what they've learned. Alternative assessment, often referred to as authentic assessment, attempts to determine if the student's knowledge is "usable." The Rivers Project curriculum emphasizes authentic assessment tools. Authentic assessment assignments usually do not have a single answer, and students must employ critical thinking skills and creativity. Two of the more popular forms of authentic assessment are performance assessments and portfolio assessments. In both, students demonstrate how they use the classroom instruction to find a logical solution to a problem.

Many of the concluding activities for specific lessons work well as performance assessments. Most include performance criteria and a scoring rubric. Teachers may also use their own performance criteria and scoring rubric for such standard components of geography classes as data handling, map reading, map drawing, and written analysis. Appendix A contains two performance-based activities designed as unit-end assessments.

Portfolio Assessment

An academic portfolio is not an accumulation of all the assignments and tests of a grading period; rather, it is a collection of selected documents organized as evidence of a student's cumulative learning. Like an artist's or model's portfolio, the academic portfolio is an opportunity for the students to demonstrate the depth and breadth of their learning. The student takes responsibility for selecting what to include in the portfolio and how to organize it. Students must feel a sense of ownership for their portfolios. The goal of the portfolio process is that the student reflects on his or her personal learning process and recognizes learning as a process that extends beyond textbooks and lectures. Teacher Notes and a student handout on portfolios are included in Appendix A.

Selecting Assessments

Most *Rivers Geography* activities work well either as ungraded tasks, evaluated simply for participation or completion, or as more formally evaluated assessments. From the chart following, select the assessment tools that suit your preferences and circumstances, or devise your own.

Assessment	Assessment Type	Location
Student Activities		
1.5 A River Collage	performance	end Lesson 1
2.3 Investigating the Location of Watersheds	forced-choice	end Lesson 2
3.6 Sketching a Map	performance	end Lesson 3
4.2 Tracing the History of a Local Family	performance and forced-choice	end Lesson 4
5.3 Simulation of Environmental Decisionmaking	performance and forced-choice	end Lesson 5
6.3 River or Stream Field Study	mostly performance	end Lesson 6
7.2 River or Stream Brochure	performance	end Lesson 7
Unit-End Assessments		
Portfolio	performance	Appendix A
Unit-end Test	forced-choice	Appendix A
Analyzing Geographic Information	performance	Appendix A
Challenge Project: Telecommunications	performance	Appendix A
Unit-Long Assessments		
Notebook	performance	Lesson 1
Journal	performance	Lesson 1

TEACHER NOTES

How to Begin a River Study

Focus: The Five Themes of Geography

In this first lesson, students will become aware that the Earth's supply of fresh water is limited and endangered. They will learn to relate the five themes of geography to a study about river systems. They will analyze an article about a river problem or controversy. Finally, they will express their ideas and thoughts about rivers by writing in their journals and by making a river collage.

Place
Location
Region
Movement
Human-Environmental Interaction

Perspectives and Geography Standards

Perspectives: Spatial and Ecological

Geography Standards Receiving Primary Emphasis

This lesson primarily promotes mastery of the following standards:

1. How to use maps and other geographic representations, tools, and technologies to acquire, process, and report information from a spatial perspective
3. How to analyze the spatial organization of people, places, and environments on Earth's surface
14. How human actions modify the physical environment

Geography Standards Receiving Secondary Emphasis

Several other standards receive secondary emphasis during specific activities. Specifically, student journal entries and the analysis of an article about a river problem encourage the use of mental maps (Standard 2). The segments on freshwater and on the hydrological cycle focus on physical processes (Standard 7).

Learner Outcomes

Students will:

1. Learn that water on Earth is a finite resource that is sometimes misused.
2. Learn about the hydrologic cycle.
3. Categorize information according to the five themes and identify examples of the five themes in illustrations.

4. Read an article about a river and analyze the river's characteristics and problems.
5. Begin a project journal by writing an initial entry and make a river collage.

Time

Five class periods of 40–50 minutes per period (six class periods, if students are to do library research during class time).

DAY 1: Teacher Presentation 1.1: Freshwater Demonstration
Student Information 1.1: The Hydrologic Cycle

DAY 2: Teacher Presentation 1.2: Geography's Five Themes
Student Information 1.2: Your Rivers Geography Notebook
Student Activity 1.3: Keeping a Journal
Assign Student Information 4.1: Genealogy Research and Community History and Student Activity 4.2: Tracing the History of a Local Family

DAY 3: Student Activity 1.4: Analyzing Articles About Rivers

DAY 4: Student Assessment 1.5: A River Collage

DAY 5: Student presentation of collages
Assign as homework Student Information 2.1: River Systems and Student Information 2.2: Map Scale, Direction, and Location

Advance Preparation

Prepare to supply students with Student Information 1.1, 1.3, 2.1, 2.2, and 4.1; Student Activity 1.2, 1.4, and 4.2; Student Assessment 1.5; the Glossary; and base maps of your country and of the world. (A base map shows general information, such as outlines of land features, rivers, or political boundaries. For a source of base maps, see National Council for Geographic Education, in Appendix C).

Decide whether to include portfolios as project-long assessment tools, as presented in Appendix A. If so, prepare to supply students with Student Information sheets about portfolios now or near the end of the unit.

Gather all necessary materials for this lesson. Make arrangements for each student to have a journal for the entire unit. Each student will also need a looseleaf binder to hold materials distributed throughout the *Rivers Geography* unit. Either inform students that they should bring these, or arrange to provide them. If you wish, display a poster or transparency of the hydrologic cycle. Select slides, photographs, or pieces of literature to use in a classroom discussion about the five themes.

Read Lesson 1 and become familiar with the geographic concepts that apply to the lesson. (Excellent resources on geographic themes and standards are mentioned in the introduction on page xvi and listed in Appendix C.)

Practice Teacher Presentation 1.1: Freshwater Demonstration.

Schedule library time for students to find articles about rivers. If you do not wish students to find their own articles, locate and photocopy readings for your class. A particularly good source is Social Issues Research Services, Inc. (SIRS), which reprints articles from a variety of periodicals and arranges them in volumes according to subject matter.

Materials

Teacher Presentation 1.1: Freshwater Demonstration

For the teacher

4 or 5 ice cubes

measuring cup, ½-cup (4-oz) capacity

1-gallon container of water

4 1-quart containers or 16 8-ounce glasses

measuring tablespoon

plastic syringe or eyedropper

paper towels

Teacher Presentation 1.2: Geography's Five Themes

Per two students

poster entitled *Maps, the Landscape, and Fundamental Themes of Geography* (from National Geographic Society)

For the teacher

slides or photographs illustrating the five themes of geography

Student Activity 1.2: Your Rivers Geography Notebook

Per student

three-ring notebook

lined three-hole paper

unlined three-hole paper

8 tabbed dividers for notebook

Student Activity 1.3: Keeping a Journal

Per student

notebook, bound so pages cannot be removed or inserted (unless teacher prefers or accepts spiral-bound notebook)

Student Activity 1.4: Analyzing Articles About Rivers

Per student

magazine article (or feature-length newspaper article) about a river issue

atlas

Per group

dictionary

Optional

computer with CD-ROM drive

CD-ROM disc of magazine articles

Student Assessment 1.5: A River Collage

Per student

poster board (½ sheet)

scissors
glue stick
Per class
old magazines from which pictures and text can be taken
markers
colored paper

Vocabulary

atmosphere
condensation
evaporation
five themes
fresh water
glacier
groundwater
human-environmental interaction
hydrologic cycle
ice cap
journal
location
movement
perspective
place
pollution
precipitation
region
runoff
soil moisture
transpiration

Background for the Teacher

No one can understand an entire river system by observing just one location. Similarly, before investigating a specific portion of the local stream or river, students need an overall picture. Individual studies take on greater significance when students understand their relationship to the entire system. Geography's five themes—Location, Place, Movement, Human-Environmental Interaction, and Region—familiar to most geography teachers, help students do this. (Any teacher without a working knowledge of the five themes should study not only the material in Teacher Presentation 1.2: Geography's Five Themes but also some of the excellent publications listed in Appendix C.)

Just as the streams that feed a larger river give structure to that river's watershed, the individual activities contained in *Rivers Geography* give shape to the entire unit of study. The presentation of geography's five themes in this first lesson gives you a way to help students understand the overall organization of this unit. The grid on pages xiv and xv of the unit introduction connects each lesson in *Rivers Geography* with specific questions pertinent to each theme.

Introducing the Lesson

1. Perform for your class Teacher Presentation 1.1: Freshwater Demonstration. Encourage students to discuss and record the various uses of water in their daily lives.
2. List on the chalkboard the vocabulary terms *(fresh water, soil moisture, groundwater, pollution, ice cap, glacier,* and *atmosphere)* presented in the demonstration. Distribute a copy of the Glossary to each student and have

students look up the definitions of new terms. Discuss the vocabulary with the class.

3. If you wish to assess students' prior knowledge of the water cycle, ask them to draw a labeled diagram of the water cycle before handing out Student Activity 1.1: The Hydrologic Cycle. Then have students read Student Activity 1.1 and answer the questions. Discuss with students how the river is an important component of the world's freshwater resources.
4. Relate the hydrologic cycle to geographical features in your locale. Display a poster or overhead transparency of the hydrologic cycle, or sketch the cycle on the chalkboard. Have students describe or draw the hydrologic cycle. For additional specific teaching activities on the hydrologic cycle, see *Rivers Earth Science*.

Developing the Lesson

1. Give students an overview of the *Rivers Geography* unit. Discuss with students the nature of the lessons and activities they will be doing. If appropriate, point out that students will have opportunities via the Internet to share the results of their work with hundreds of schools participating in the Rivers Project.
2. Discuss the five themes of geography, as detailed in Teacher Presentation 1.2: Geography's Five Themes. Begin by distributing to pairs of students the National Geographic Society poster titled *Maps, the Landscape, and Fundamental Themes of Geography.* (Sharing these materials in pairs gives students a good opportunity for cooperative learning. These posters are also quite large, so they fit better on two desks together than on a single desk.)
3. Have students read Student Activity 1.2: Your Rivers Geography Notebook and Student Activity 1.3: Keeping a Journal. Pass out any materials you are providing. (Though these handouts indicate students should take class notes on loose-leaf paper, alter these instructions if desired.) Answer any questions students may have, making sure they understand the different purposes of the notebook and the journal. Have them write a journal entry based on their perceptions of the introductory activities. Encourage students to share their thoughts regarding the significance of water in their daily lives and their relationships to the river or stream you are about to study.

   If students are to provide their own journals and do not yet have them, instruct them to write their entries in their class notebooks. They can transfer the entries to their journals later.

4. Distribute and discuss with students information on any other unit-long assessment tools you plan to use, such as portfolios. (You may distributed the detailed handout about portfolios in either Lesson 1 or Lesson 7.)

5. Distribute and discuss with students Student Information 4.1: Genealogy Research and Community History and Student Activity 4.2: Tracing the History of a Local Family. Distribute base maps for your country and for the world. Assign students to complete Student Information 4.1, and to complete Part A of Student Activity 4.2 (the family tree). Explain that you are passing out these materials early so students will have ample time to do research and gather data. Make sure students clearly understand the purpose and process of building a family tree. Be sensitive to students who may be adopted or living in foster homes. If any students are not comfortable researching their own families, make sure they know that they may research the genealogy of a friend, neighbor, or some other member of the community. (For further background information on this activity, see the corresponding Teacher Guide for Student Activity 4.2.)

Concluding the Lesson

1. Have students perform Student Activity 1.4: Analyzing Articles About Rivers. Provide information about potential topics, article length, and possible sources of articles, then permit library research time for students to select articles. Alternatively, you may select, photocopy, and distribute articles to students. Depending on the maturity of your students and their familiarity with the library, you may need to take the class to the library to conduct a search for articles.
2. The next day, have students share their reactions to the articles in small groups. Then discuss as a class the articles, questions, and student responses.

Assessing the Lesson

1. Read through with the class Student Assessment 1.5: A River Collage. Instruct students to work individually or in pairs to create their own river collages. You may provide students with basic materials, but encourage students to include other materials if they wish. (Old magazines are occasionally discarded by the library, or they can be brought from home.) The collages may be done partly or entirely as homework. By providing the students with the opportunity to express themselves about a river or stream topic, the collage activity should generate student interest, involvement, and commitment. Furthermore, the collage activity requires several higher-order thinking skills: conceptual organization, classification of information, self-analysis, critical review, and peer review.
2. Display completed collages on a wall or bulletin board. Each student or pair of students briefly describe to the class the goal of the collage and how the goal was achieved. Applause for students' presentations or a snapshot of students posing with their collages can provide positive reinforcement and inspire enthusiasm about *Rivers Geography*.

3. Assess collages. Commitment and honest effort should be more significant than artistic merit. The suggested performance criteria are included in Student Assessment 1.5. Here is a suggested scoring rubric to use with those criteria:

Scoring Rubric

Score	**Expectations**
0	No response attempted.
1	Little or no attention to criteria; message absent or obscure; focus on a river or stream topic absent or obscure; materials are unrelated; minimal variety of materials, visually unappealing.
2	Limited attention to criteria; message stated but unclear; focus on a river or stream topic uncertain; materials limited in variety and effectiveness, limited visual appeal.
3	Satisfactory attention to criteria; message understandable; clear focus on a river or stream topic; generally satisfactory variety and effectiveness of materials used; visually appealing.
4	Strong attention to criteria; message strongly stated; focus on river topic unmistakable; creative and effective use of variety of well-integrated materials; visually outstanding.

Extending the Lesson

1. Though this curriculum requires only one field-site visit to your local river or stream, classes that have taken extra visits report increased enthusiasm and appreciation for the issues covered in *Rivers Geography.* If desired, during Lesson 1, take students on a field-site visit to your local river or stream. You may wish to encourage journal writing, or make observations in the form of a streamwalk, discussed under Extending the Lesson in the Teacher Notes for Lesson 6. For field-trip planning and safety information, see Lesson 6.
2. Show the National Geographic Society filmstrip entitled *More than Maps: A Look at Geography.*
3. Instruct students to list their normal daily uses of water. Supplement the lists with information from sources such as the *Student Information Kit* from America's Clean Water Foundation, 750 First Street, NE, Suite 911, Washington, DC 20002. Phone: 202-898-0902.

TEACHER

PRESENTATION

1.1

Freshwater Demonstration

Purpose

To demonstrate the fragility of Earth's freshwater supply.

Background

The demonstration is designed to focus students' attention on the overall supply of water on Earth and the relatively small supply of fresh water available for human use. Students are passive participants in this demonstration, so its effectiveness will depend on your dramatization of the material.

You may wish to have several students present the demonstration instead of doing it yourself. One student can read the narration; a second student can measure out the water.

Materials

4 or 5 ice cubes
measuring cup, ½-cup (4-oz) capacity
1-gallon container of water
4 1-quart containers or 16 8-ounce glasses
measuring tablespoon
plastic syringe or eyedropper
paper towels

Instructional Procedure

1. Arrange all the materials on a table in your classroom. Place the ice cubes in the measuring cup.
2. Present the following narration:

"Water covers 71 percent of the Earth's surface, yet it makes up only one four-thousandth (0.00025) of the Earth's total volume.

"If we let this gallon of water represent the total water on Earth, about half is the Pacific Ocean." **(Pour half the water into two 1-quart containers.)** "Most of the other half is the Atlantic, Indian, and Arctic Oceans." **(Pour all but about ½ cup into the remaining two quart-size containers.)** "Together, the oceans and saline lakes and inland seas make up more than 97 percent of Earth's water.

"The remainder is Earth's freshwater supply, which is represented by about 3½ ounces of our total gallon of 128 ounces. **(Hold the one-gallon container aloft.)** Human beings and most land and freshwater organisms depend on this freshwater supply to live."

(Pour about one ounce of water into the measuring cup filled with ice cubes.) "About three fourths of the world's fresh water—or two percent of the world's total water supply—exists in ice caps and glaciers." **(Remove the ice cubes from the cup and display them in your palm. Then drop them into one of the quart containers.)**

"About this much **(dip your fingers into the cup and flip them into the air to remove the water),** or one thousandth of a percent, is present in the atmosphere.

"All that is left is about five-eighths of one percent." **(Measure out one tablespoon of water and hold up the tablespoon.)** "The water in this tablespoon represents the total fresh water in lakes, rivers and streams, soil moisture, and groundwater.

"Some of the groundwater is very deep and is not immediately usable." **(Pour off half of the water in the tablespoon.)**

"Much of the remaining fresh water is in remote regions of the world, such as the Amazon and Siberia, and is not readily available for human use.

(Hold up the half-empty tablespoon.) "The water in this tablespoon represents our freshwater supply—the precious liquid resource that makes life possible on Earth. Remember, your body and the bodies of most living things are about 70 percent water.

"Now suppose some people decided to throw garbage into our freshwater supply. What if they dumped motor oil or gasoline into our water, flushed their toilets into it, threw old tires, empty soda cans, refrigerators, and other things into it? What if they dumped lawn pesticides, weed killers, paints, paint thinners, and all sorts of other chemicals into it, because they didn't know what else to do with them?

"Maybe a factory needs an inexpensive way to dispose of its wastes or it will go out of business. Should it be allowed to dump its wastes in our freshwater supply?

"Suppose a city has a sewage disposal system that needs updating, but it doesn't have the money? What should the city do?

"Once it was believed that rivers could simply dilute and carry away wastes, and in a much simpler society, rivers did—but no longer. The capacity of our rivers and lakes to handle wastes and still recover has been stretched to the limit and, in some cases, exceeded.

"In order to ensure our supply of fresh water, we can no longer use rivers as sewers just because it is convenient or seemingly inexpensive to do so. We have learned that we must pay for proper disposal now or end up paying much more in the future. The debts of the past are now due. The future has caught up with us." **(Place the tablespoon on the table and draw the water into an eyedropper or syringe.)**

"Each one of us must accept responsibility for safeguarding our most precious resource—our freshwater supply." **(Approach a student and indicate that the student is to hold his or her palm up. Dispense one drop on the student's palm. Continue to other students.)**

"In the United States, the average person uses between 150 to 200 gallons of clean, fresh water each day. If this average were to include all the industrial and agricultural uses of water, the daily amount would be much greater—maybe ten times more.

"People in industrialized regions, such as the United States, Canada, Europe, and Japan, put much more pressure on the environment and consume much more water than do people in undeveloped countries. Because you live in the United States, you use much more water than someone living in undeveloped parts of Africa or South America, for example." **(Dispense several more drops of water on the palms of students' hands until the water runs out.)**

(When the water runs out, say) "Sorry, there is not enough fresh water to go around. Many places in the world cannot provide citizens with adequate supplies of clean fresh water. You see, many parts of the world are already facing freshwater shortages. The fresh water that we take for granted in this country is simply not available to many other peoples of the world. Now, even in the United States, we are facing freshwater shortages because of overuse and pollution of our water resources, undesirable changes in our air, water, and land that can harm living organisms.

"The time for action is now. We, who have the capacity to influence the survival of life on Earth, must assume responsibility. The future of the world's freshwater supply—literally and figuratively—is in your hands."

TEACHER TABLE 1-1

World's Water Supply Represented By One Gallon of Water

Location	**Portion of One Gallon**
Pacific Ocean	2 quarts
Arctic, Atlantic, Indian Oceans, saline lakes, and inland seas	2 quarts minus ½ cup
Ice caps and glaciers	2¾ ounces (⅓ cup)
Atmosphere	⅛ ounce
Remainder	8/10 ounce (½ tablespoon)

TEACHER

PRESENTATION

1.2

Geography's Five Themes

Purpose

To introduce students to the **five themes of geography** and have them explore how these themes relate to the study of rivers and streams.

Background

Though the teacher should be comfortable with discussing and applying the five themes, for quick review, remember that the basic content of geography's five themes is as follows (look again at the grid on pages xiv and xv):

1. **Location: Position on the Earth's Surface** Absolute location is commonly stated by using latitude and longitude, compass heading and distance, or an alpha-numeric grid. A street address is also a form of absolute location, as is a precise written description. The Global Positioning System (GPS) uses satellites to accurately pinpoint locations. Relative location, or the location of one place or phenomenon with respect to other important places or phenomena, may be necessary to show relationships.
2. **Place: Physical and Human Characteristics** Each location on Earth is more or less distinct from all other locations, owing to the unique combination of characteristics present. Physical characteristics are produced by geological, hydrological, atmospheric, and biological processes. Human characteristics are shaped by ideas and actions and are expressed in the ways humans organize their economic, social, and political activities.
3. **Movement** Humans interact with one another through systems of transportation and communication, which facilitate the movement of people, ideas, and products. Patterns on many different scales result from the interdependence of people in various places and regions, and the nature of those relationships is subject to change. Movement also takes place in the physical world. Examples include wind systems, the hydrologic cycles, animal migrations, and the interdependence of life forms in an ecosystem.
4. **Human-Environmental Interaction** The world's population is unevenly distributed, attesting to the fact that all places have comparative advantages and disadvantages for human use. In order to serve their needs, humans adapt to the natural environment or modify it in ways that express their values and technology. Each location is unique in the way humans interact with the environment.
5. **Regions: How They Form and Change** The region is the basic study unit of geography, sometimes used as its own focus of study, sometimes used as a building block to expand knowledge about the world. A region may be large or small depending on the purpose of the study, but it must display some unity in terms of selection criteria. Investigators may define a

region by a single physical or human characteristic, such as a watershed, a political unit, a landform type, or an area occupied by people who speak a particular language, or they may examine the relationships of several or many complex characteristics represented in that region. For example, someone focusing on a watershed region might examine its rainfall pattern, its economies, or the ethnic backgrounds of the people settled there.

Materials

Per two students
poster entitled *Maps, the Landscape, and Fundamental Themes of Geography* (from National Geographic Society)
For the teacher
slides or photographs illustrating the five themes of geography

Instructional Procedure

1. Distribute to pairs of students the National Geographic Society poster titled *Maps, the Landscape, and Fundamental Themes of Geography.*
2. Have students read the explanation of each theme. Pause after each theme to discuss it and how it applies to the example provided.
3. Instruct students to write the five themes on a single page in their notebooks, leaving several lines between each theme. Present one or more appropriate slides, photos, or literature selections that demonstrate that theme in relation to a river or stream. Instruct individual students or cooperative groups of students to write under the appropriate theme examples from these illustrations.
4. Afterward, discuss students' responses. Identify and correct any misconceptions students may have about the themes.

Name ______________________ __________

The Hydrologic Cycle

What do you think of when someone says the word *water?* Do you picture a refreshing glass of drinking water? A thundering waterfall? A rushing river or stream? A shimmering lake? A vast expanse of ocean? It is likely that when you think of water, you picture it in its liquid state instead of in its solid state (ice) or its gaseous state (water vapor).

The Three States of Water

Over 97 percent of Earth's water is liquid—mostly salt water filling oceans, inland seas, and saline lakes. Less than one percent of Earth's liquid water occurs as **fresh water,** or water that has a low salt content and is, therefore, drinkable. Without fresh water, most life on Earth could not exist.

When you fill an ice-cube tray with tap water and place it in the freezer, the liquid turns to solid ice. So, too, in the coldest regions of Earth—near the poles or the tops of mountains—about two percent of Earth's water occurs in a solid state as ice caps and glaciers. An **ice cap** is a large, permanent mass of ice that generally extends over land surfaces but may project over adjacent water areas as well. A **glacier** (GLAY sher) is a large body of perennial ice that moves slowly down a slope.

When you boil tap water, some of the liquid vaporizes, changing into a gaseous state. In nature, too, a small amount of water (one-thousandth of a percent) occurs as water vapor in the lower **atmosphere,** the thin layer of gases enveloping Earth.

TABLE 1-1

Distribution of Water on Earth

Location	Percentage of Total Water on Earth
Oceans	97.2
Ice caps and glaciers	2.15
Groundwater	0.625
Freshwater lakes	0.009
Saline lakes and inland seas	0.008
Atmosphere (at sea level)	0.001
Rivers and streams	0.0001
Total (rounded)	100.0000

Source: Adapted from *Water of the World,* U.S. Geological Survey

The Water Cycle

The water on Earth is constantly changing from one state to another. When warmed, ice and snow melt into liquid. If the temperature increases enough, water changes into water vapor, which rises into the lower atmosphere. When cooled, water vapor in the atmosphere changes into rain, sleet, or snow that falls to the surface of the earth, starting the cycle all over again. The endless cycling or interchange of water among the oceans, the lower atmosphere, the land surface, and reserves several kilometers below the earth's surface is called the water cycle, or **hydrologic** (hi druh LAHJ ihk) **cycle.**

As water absorbs the sun's energy, it vaporizes, or changes into a gaseous state, by the process of **evaporation** (ih vap uh RAY shuhn). As water vapor cools, it releases energy and changes back into a liquid state by the process of **condensation** (kahn dehn SAY shuhn). Dew forms when water vapor condenses on the land surface. Rain or liquid **precipitation** (prih sihp uh TAY shuhn) occurs when water condenses in the atmosphere, depositing water that reaches the earth's surface. If atmospheric temperatures are cold enough, the raindrops may lose more heat, turning into snow or ice.

Gravity, the rotation of Earth, and the uneven heating and cooling of Earth combine to drive the hydrologic cycle. Evaporation from the oceans contributes some 86 percent of atmospheric moisture. The remaining 14 percent comes from evaporation from land surfaces and transpiration by plants. **Transpiration** (trans puh RAY shuhn) is the process by which moisture in plants is changed to vapor.

Condensation results in precipitation, of which 78 percent falls on water surfaces and 22 percent falls on land surfaces. Precipitation and **runoff** (the lateral movement of surface and ground water) fills lakes, rivers, and streams. Water seeping into the ground replenishes **soil moisture**—water in the soil layer that is available to plants— and **groundwater**—the water that collects beneath the surface of the earth. Although rivers and streams account for only 0.0001 percent of the world's total water supply, they play a vital role in the hydrologic cycle and in the lives of humans as well.

DRIFTWOOD

"The balance of nature is not a *status quo;* it is fluid, ever shifting, in a constant state of adjustment."

From *Silent Spring* by Rachel Carson

DRIFTWOOD

"We are part of this earth and it is part of us."

Chief Seattle

Questions

Write your answers on a separate piece of paper. Use complete sentences.

1. What are the three states in which water exists on Earth? Which is the most common?
2. In your own words, describe the hydrologic cycle.
3. Where is most of the world's water located?
4. What is the river's role in the hydrologic cycle?
5. How do humans use the river?
6. Give some examples of human activities that contribute to **pollution** of our water resources, undesirable changes in air, water, and land that can harm living organisms.

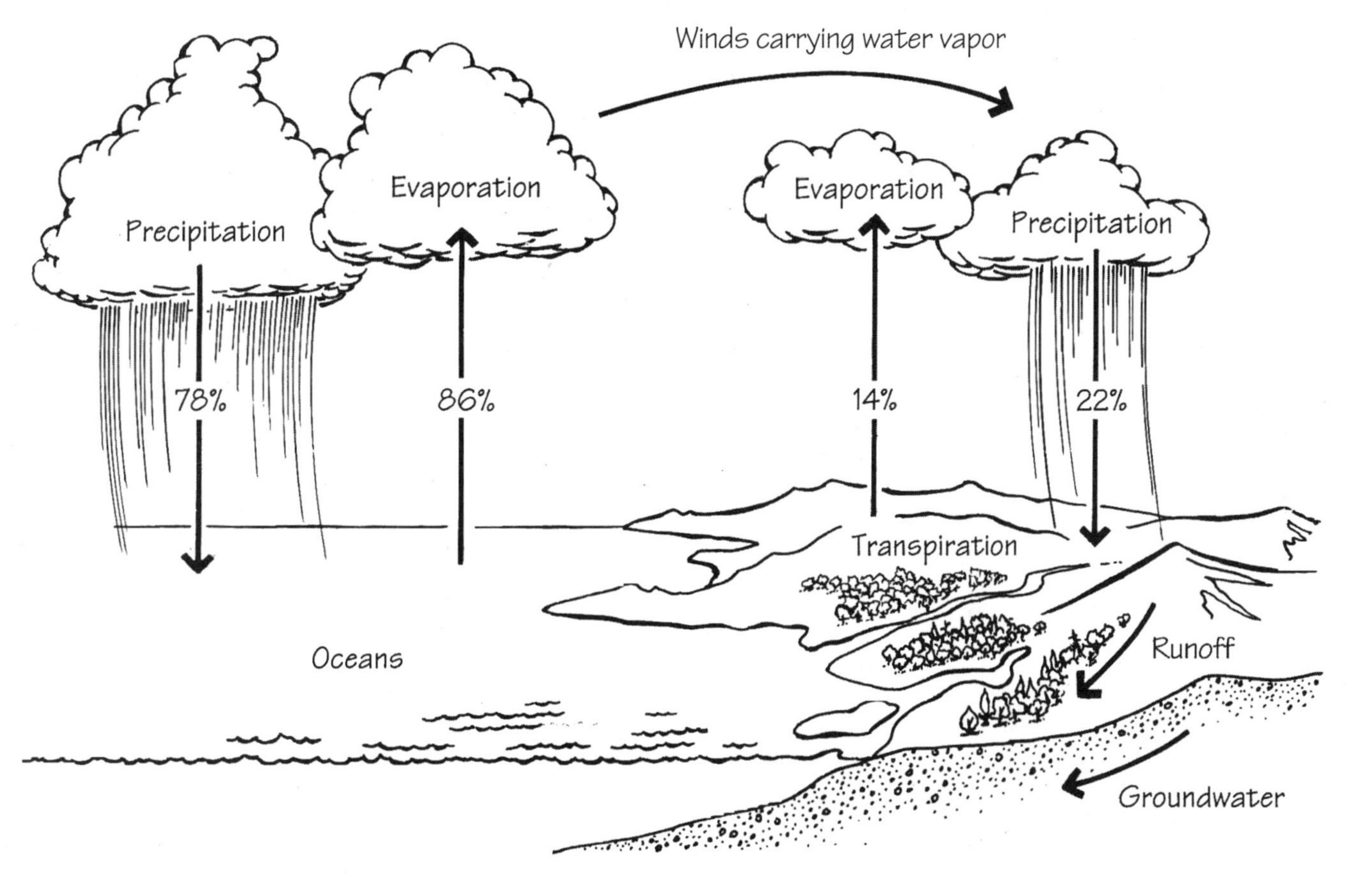

Figure 1: The Hydrologic Cycle

Name

STUDENT ACTIVITY 1.2

Your Rivers Geography Notebook

Purpose

To organize the materials you receive and create during the *Rivers Geography* unit.

Background

Throughout the *Rivers Geography* unit, you will receive information sheets, activity sheets, and assessments in the form of photocopied handouts. For many of these, you will also have to write answers or make sketches. You will also need to take notes in class, when you visit your local river or stream, at the library, and in interviews with local individuals. In order to help organize this information, you should keep it in a separate three-ring notebook specifically reserved for *Rivers Geography*. (As described in another sheet, you will also keep a journal in which you record your personal thoughts and reflections.)

Materials

Per student

- three-ring notebook
- lined three-hole paper
- unlined three-hole paper
- 8 tabbed dividers for notebook

A well-kept notebook is a measure of your progress in the unit. You will find it an indispensable source of information for writing answers to questions, preparing for tests, doing unit projects, and writing journal entries. Your teacher may ask to look at your notebook from time to time to assess your progress. Be sure your notebook is up-to-date for each class meeting.

Procedure

1. Label and insert an index tab for each of 7 lessons in *Rivers Geography,* plus one for "Notes."
2. Write your name in the front of your notebook, along with "Rivers Geography."
3. Keep a supply of loose-leafed lined paper in your notebook. Write answers to questions and other assignments on this paper.
4. Keep unlined paper in your notebook for making maps and other sketches.
5. Keep each written set of answers or other responses to assignments in your binder just behind the corresponding assignment sheet, behind the tab for the appropriate lesson. Label each assignment with your name, date, and the assignment name.
6. Keep your class notes on paper separate from your answers to assignments. Make an entry for each class meeting. Date each entry. Skip several lines between entries to separate them clearly.
7. During classroom activities, record observations, data, and ideas in as great detail as time permits. If you have time at the end of class, write a brief summary of the class session. If not, be sure to summarize before the next class meeting.
8. Keep your class notes behind the tab labeled "Notes."
9. In your notebook, record any field observations, interviews, outside research, and other ideas and questions about the project that occur to you outside class.
10. When you take notes in other settings, such as during a site visit or interview, be sure to label each page with the date, as well as location or name of individual, as appropriate.
11. Bring your notebook to all class meetings, and keep it handy outside class.

Performance Criteria

- Notebook is up-to-date.
- Notebook is organized as prescribed.
- Each entry is clearly separated from others, dated, and labeled appropriately.
- Each entry pertaining to class activities contains a body of notes and summary as appropriate.

Name

Keeping a Journal

Purpose

To create an organized way to record your thoughts and reflections on the issues you encounter throughout *Rivers Geography*.

Background

Professional geographers and others who conduct studies must keep records of not only their activities and observations but their thoughts in order to gain a thorough understanding of their topics of research. They know that such a **journal** helps them organize their thoughts and understand more clearly how all the parts of their study fit together. In this way, their studies become more than collections of facts. The journal helps them achieve a deeper, more complete understanding of their research topic.

The journal that you will keep during *Rivers Geography* should contain your impressions of class activities, your thoughts about the river or stream you are studying, and a record of your experiences. Journal entries may include impressions, summaries, and critiques of class activities, teacher-assigned topics, or topics that you select. Your journal should help you:

- Organize your thoughts.
- Make sense of your observations.
- Think more clearly and effectively.
- Learn content material more easily.
- Improve your writing skills.
- Learn more about yourself

Materials

Per student

- Notebook, bound so pages cannot be removed or inserted (unless teacher prefers or accepts spiral-bound notebook.)

Procedure

1. Write your name in the front of your bound notebook, along with "Rivers Geography."
2. Make at least two journal entries per week.
3. Begin each entry on a new journal page.
4. Begin each entry with the day of the week, the date, location, and a title that describes the content of the entry.
5. Keep your journal separate from other notes and class materials so that it can be collected periodically by your teacher.
6. Always bring your journal to class, and keep it handy outside of class so you can jot down notes about thoughts that occur to you. Always take it when you go on field-site visits.
7. Write all journal entries legibly in ink (except sketches).
8. When appropriate, draw a sketch map to accompany your entry.

Performance Criteria

The grade value of your journal will equal _____ percent of your total grade for the unit. The journal will be collected for assessment at the conclusion of the *Rivers Geography* unit. The successful journal:

- Shows clarity of focus about the concept to which specific entries relate.
- Includes at least two entries per week.
- Demonstrates understanding of geographic ideas.
- Makes valid applications of geographic concepts to the real world.
- Supports a focused concept through reasoning, detailed observations, and documentation.
- Uses correct, standard written English.

Name

Analyzing Articles About Rivers

Purpose

To learn how to read and analyze magazine articles about a river problem or controversy.

Background

Part of the challenge of research is finding appropriate sources of information. The *Reader's Guide to Periodical Literature* indexes magazine articles and is a logical starting point for your investigation of a river problem. Your library may have back issues of magazines in print, on microfilm, or on CD-ROM.

Reprinted articles on a variety of topics are indexed by subject matter and contained in *Social Issues Resource Series* (SIRS). Articles also may be found in your library's vertical files. If you do not have enough time to read your article in the library or during class, perhaps you can photocopy the article and finish this activity at home.

Materials

Per student

- magazine article (or feature-length newspaper article) about a river issue
- atlas

Per group

- dictionary

Optional

- computer with CD-ROM drive
- CD-ROM disc of magazine articles

Procedure

1. Select a magazine article that describes a problem of humans interacting with a river or some other natural water feature.
2. To establish an analytical framework in your mind, read the questions in this activity before you read the article. The questions in this activity are designed to help you analyze the magazine article.
3. To answer all the questions, you may need to read the article several times. Keep a dictionary handy to look up words you do not understand.
4. Write your answers to all questions on a separate piece of paper. Use complete sentences.

Observations

1. Provide a bibliographical reference for the article by listing the author or authors (last name first for the lead author), title of the article in quotation marks, title of the magazine underlined, date of publication, pages (first to last page of article).
2. In one sentence, state the problem or controversy addressed in the article.
   a. Is the problem or controversy due to a decision that has already been made or a decision that will be made?
   b. Who made or will make the decision?
   c. What were or are the choices?

Analyses and Conclusions

1. Is the title descriptive of the article's content? Suggest another title.
2. Does the author favor one point of view, or does the author present all sides of the issue? Cite examples to support your answer.
3. List the sources the author used to obtain information. If sources are not stated, speculate where the author obtained the information to write the article?

4. Explain how human activities have altered the river or other body of water described in the article. List the changes and explain the effects of each.
5. Geographers commonly look at topics from a spatial perspective by identifying where phenomena (facts or events) exist and why they exist there. In general, a **perspective** is a point of view that provides a frame of reference from which to ask questions; acquire, organize and analyze information; and answer questions. Specifically, a spatial perspective enables geographers to understand how several phenomena exist and influence one another at the same location or in the same region. Based on the article, write a statement to explain the relationship of the human society to the natural river environment.
6. Additional perspectives may aid your understanding of a particular topic. See if the following perspectives are present in the article.
   a. Historical perspective adds the dimension of time. What events led to the current condition? When and why did these events occur? How did these events change conditions when they occurred? What impact are these events having now? Explain the article's problem or controversy from a historical perspective.
   b. Economic perspective concerns the ways people produce, exchange, and consume goods and services. Because economies are interrelated, it may be necessary to look at economic activities on a regional, national, or even international scale. Earning a living may involve altering the physical environment to accommodate economic needs, such as developing natural resources and building roads and communication lines. Write a statement to explain the river problem from an economic perspective.
7. In one paragraph (3 to 6 sentences), summarize your article.

Critical Thinking Questions

1. On a separate piece of paper, sketch a map of the river or other body of water that is the focus of your article. Plot the locations of places or events mentioned in the article. Consult an atlas to find additional information. Plot and label cities, landforms, land uses, political boundaries, and other topics in the article.
2. What are the major similarities and differences between the river in the article and your local river or stream?
3. What are the problems or controversies involving your local river or stream? How are they being handled?

Keeping Your Journal

Express additional thoughts or ideas in your journal. Here are some suggestions:

1. How are rivers valued by humans?
2. How can one's opinion about the local river be influenced by reading about other rivers in other places?
3. How can constructing a sketch map of the river in the magazine article aid your understanding of the river's problems?

Name

A River Collage

Purpose

To present a message about a river or stream by constructing a collage.

Background

A collage is a collection of photos and other materials, arranged in order to convey some message to the viewer. The collage should be visually pleasing and should invite some response from the viewer. Color, form, and the selective use of white space contribute to the effectiveness of a collage.

Before making a collage about your river or stream, you will need to decide what message you wish to convey with your collage. Then you will need to select and arrange items that most effectively accomplish your goal.

What impressions of the waterway would you like to express to others? The beauty of the river or stream? Its value as habitat? Its value to humans as a water supply, a source of transportation, or a recreational resource? Its power and destructiveness in a flood situation? Perhaps you want to focus on human-made problems caused by the removal of vegetation along the banks of the river or stream, construction of dams, drainage of wetlands, dredging, pollution, or overuse by industries or municipalities.

Materials

Per student

- poster board (½ sheet)
- scissors
- glue stick

Per class

- old magazines from which pictures and text can be taken
- markers
- colored paper

Procedure

1. Decide on a topic related to rivers or streams that you can portray visually.
2. Decide what message you want to convey.
3. Decide whether your collage will need verbal components. If so, create these.
4. Select and gather materials. Depending on how you envision the message you want to present, you may want to use materials in addition to those listed, such as yarn, a portion of a map, fabric, or natural objects. Use the performance criteria at the end of this handout to guide your work.
5. Experiment with the sizes, shapes, colors, and placement of materials on the poster board to achieve maximum impact.
6. Glue the materials and other components to the poster board.
7. Write answers to the accompanying questions on a separate piece of paper. Use complete sentences.

Observations

1. What is the topic of your collage?
2. What message do you want to convey?
3. What materials have you used?
4. Did all the materials you selected contribute to the point that you wanted to make? Explain why or why not.
5. What features of your collage pleased you the most?
6. What changes could improve your collage?

Analyses and Conclusions

Answer these questions after you have presented your collage to others.

1. When you showed your collage to others, was your message clear? What elements of the collage were especially helpful in conveying your message?
2. In addition to your intended message, what other messages did your collage convey to viewers? How did viewers interpret your collage?

Critical Thinking Questions

1. What can a collage convey better than other forms of communication? What is difficult to convey with a collage?
2. What are some other ideas about your river or stream that occurred to you as you were constructing your collage?

Keeping Your Journal

Respond to one or more of the following statements in your journal.

1. Creating a collage was a challenge, but it didn't seem like work.
2. I was impressed with the collage presented by (a classmate) because ___.
3. The collage activity helped me organize my thoughts.
4. The collage activity helped me express what I felt.
5. The activity made me aware of some additional issues about the river or stream.

Performance Criteria

The successful collage:

- Presents a clear message.
- Emphasizes a river or stream topic.
- Integrates materials to contribute to the effectiveness of the presentation.
- Uses a variety of materials.
- Is visually appealing.

Where Is the River?

Focus: Location

In this lesson, students will develop ways to describe locations of phenomena and understand relationships that occur because of relative location. In terms of rivers, they will become acquainted with basic river vocabulary, map use, and relationships between climate and water flow.

Place
Location
Region
Movement
Human-Environmental Interaction

Perspective and Geography Standards

Perspective: Spatial

Geography Standards Receiving Primary Emphasis

This lesson primarily promotes mastery of the following standards:

1. How to use maps and other geographic representations, tools, and technologies to acquire, process, and report information from a spatial perspective
3. How to analyze the spatial organization of people, places, and environments on Earth's surface
7. The physical processes that shape the patterns of Earth's surface

Geography Standards Receiving Secondary Emphasis

References to physical and human place characteristics (Standard 4) appear throughout the lesson, and student identification of watersheds promotes the concept of region (Standard 5).

Learner Outcomes

Students will:

1. Learn what a river system is and how precipitation and evapotranspiration influence the character of a river.
2. Be able to describe the relative and absolute locations of rivers and related phenomena and to calculate distances and areas using a map scale.
3. Use atlases and maps to learn about the impact of location and other factors on the characteristics of rivers in other parts of the world.
4. Be able to identify the location of a local watershed in several different ways.

5. Compare river volumes and discharges, watershed sizes, precipitation, and temperatures and recognize the relationships among these different factors.

Time

Four class periods of 40–50 minutes per period

BEFORE STARTING: Assign Student Information 2.1 and 2.2 as homework

DAY 1: Student Information 2.1: River Systems
Student Information 2.2: Map Scale, Direction, and Location

DAY 2: Teacher Presentation 2.1: Case Study of the Spoon River Watershed
Student Activity 2.3: Investigating the Location of Watersheds, Observations (with corresponding Teacher Guide)

DAY 3: Student Activity 2.3: Investigating the Location of Watersheds, Analyses and Conclusions

DAY 4: Student Activity 2.3: Investigating the Location of Watersheds, Critical Thinking Questions
Student Assessment 2.4: Relating Location to Characteristics of Rivers
Assign Student Information 3.1 as homework

Advance Preparation

Read the lesson. Prepare to supply students with Student Information 2.1 and 2.2, Student Activity 2.3, and Student Assessment 2.4. Obtain all necessary materials and equipment. If students have access to an atlas with full-color maps of world temperature and precipitation ranges, they can answer additional questions on Student Activity 2.3. (A list of such atlases is included in the Teacher Guide designed for use with this student activity.) Prepare all overhead transparencies and other materials for Teacher Presentation 2.1 and for the Teacher Guide for Student Activity 2.3, as well as a transparency of Teacher Figure 2-1, Relative Location of Abadan, located at the end of this Teacher Notes, for use with Student Information 2.2.

If you plan to include the activities suggested in Extending the Lesson, contact your state office of the USGS and the National Climatic Data Center to obtain the necessary publications, and prepare copies of the blank climograph form, Teacher Figure 2-2, included at the end of the Teacher Notes.

Distribute Student Information 2.1: River Systems and Student Information 2.2: Map Scale, Direction, and Location as homework if not done at end of Lesson 1.

Materials

Teacher Presentation 2.1: Case Study of the Spoon River Watershed

For the teacher

transparencies of Spoon River Watershed; Estimation of Spoon River Watershed Area; Discharge of Spoon River at London Mills; and Climograph for Peoria, Illinois (Teacher Figures 2-3 through 2-6)
overhead projector
wall map of Illinois or of the United States
washable markers, several colors
ruler

Student Information 2.2: Map Scale, Direction, and Location

For the teacher

transparency of Relative Location of Abadan (Teacher Figure 2-1)
overhead projector

Teacher Guide for Student Activity 2.3: Investigating the Location of Watersheds

For the teacher

transparency of Largest World Rivers by Discharge Volumes at Their Mouths (Teacher Figure 2-7)
unlabeled map of North American rivers (Figure 2-6)
overhead projector

Student Activity 2.3: Investigating the Location of Watersheds

Per student

photocopied portion of state highway map or state geological survey map showing project watershed, or state highway map
atlas
markers and highlighters
ruler
pencil
calculator (optional)

Per class

additional relevant atlases and maps (particularly atlases with world temperature and precipitation maps)

Vocabulary

absolute location
alluvium
bank
bed
cardinal directions
channel
climograph
compass direction
course
data
delta
deposition
discharge
divide
drainage basin
equator
evapotranspiration
exotic river
floodplain
geographic grid
graphic scale
hemisphere
hydroelectricity
internal drainage
intermediate directions
latitude
longitude
map
map scale
meridian
minute
monsoon
mouth
parallel
pattern
phenomenon
prime meridian
relative location
representative fraction
reservoir
river
river system
saline lake
salinization
second
source
spatial distribution
spring
stated scale
stream
tributary
watershed
wet monsoon
wind

Introducing the Lesson

1. Give students a brief overview of Lesson 2. Discuss and review student answers for Student Information 2.1: River Systems and Student Information 2.2: Map Scale, Direction, and Location, which were assigned as homework prior to beginning this lesson. (Sample answers are in Appendix B). Have students use atlases or wall maps to locate the place names mentioned in Student Information 2.1. Also use the maps to provide examples of scale, direction, and location. Have students compare their sketch maps of the relative location of Abadan (Student Information 2.2) with your display of the transparency of Teacher Figure 2-1, located at the end of this Teacher Notes.
2. Present Teacher Presentation 2.1: Case Study of the Spoon River Watershed. Present the case study as interactively as possible so students have an opportunity to apply new concepts.

Developing the Lesson

1. Have students carry out Student Activity 2.3: Investigating the Location of Watersheds. (Corresponding Teacher Guide provides specific teaching steps and sample answers for use in parallel with Student Activity 2.3.) As they work, stop at appropriate times to discuss as a class the steps students have just completed.

Concluding the Lesson

1. Have students complete the Critical Thinking Questions in Student Activity 2.3. Discuss their answers.

Assessing the Lesson

1. Distribute Student Assessment 2.4: Relating Location to Characteristics of Rivers, with accompanying temperature-precipitation graphs and world map.
2. If you plan to collect and grade this assessment activity, have students work individually. If you want to use the exercise as a general evaluation of class progress, students may work cooperatively. Encourage them to justify decisions to one another. In either case, after students have completed this assessment, hold a classroom discussion that focuses on students' reasoning behind their decisions.

Extending the Lesson

1. Using copies of the blank climograph form at the end of the Teacher Notes for Lesson 2 (Teacher Figure 2-2), have students construct temperature-precipitation graphs of data from a weather station in or near your community. The publication, *Comparative Climatic Data for the United States through 1993,* contains temperature, precipitation, and other atmospheric data. You can obtain it by contacting the National Climatic Data Center, Federal Building, Asheville, NC 28801.
2. Have students construct graphs to compare water quality with discharge volume for various gauging stations throughout their state or river watershed. The USGS collects detailed information about all the watersheds for each state and publishes the information in *Water Resources Data* for each "water year," which runs from October to September. Contact your state office of the USGS to obtain a copy of the most recent data.

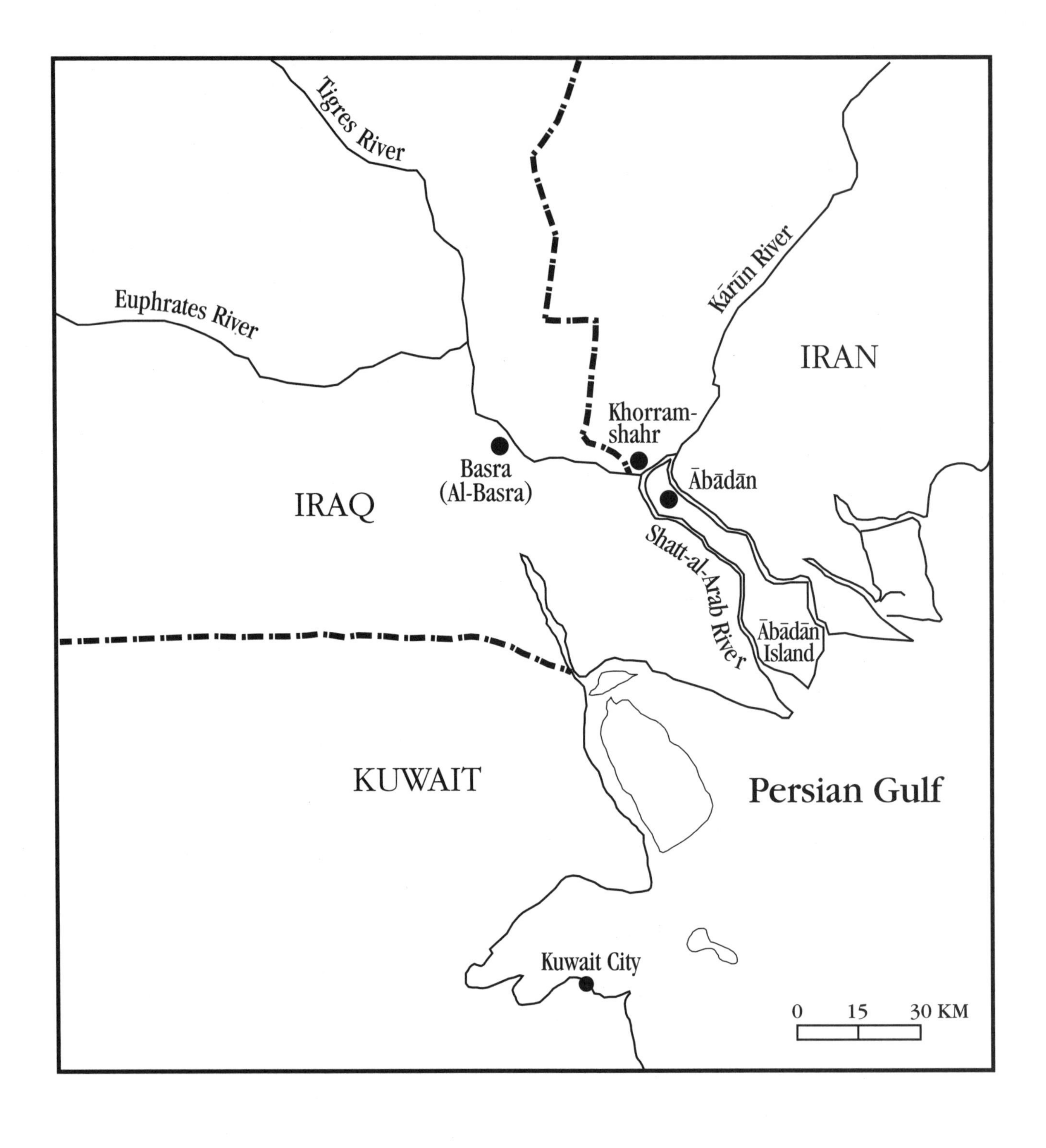

Teacher Figure 2-1: Relative Location of Ābādān

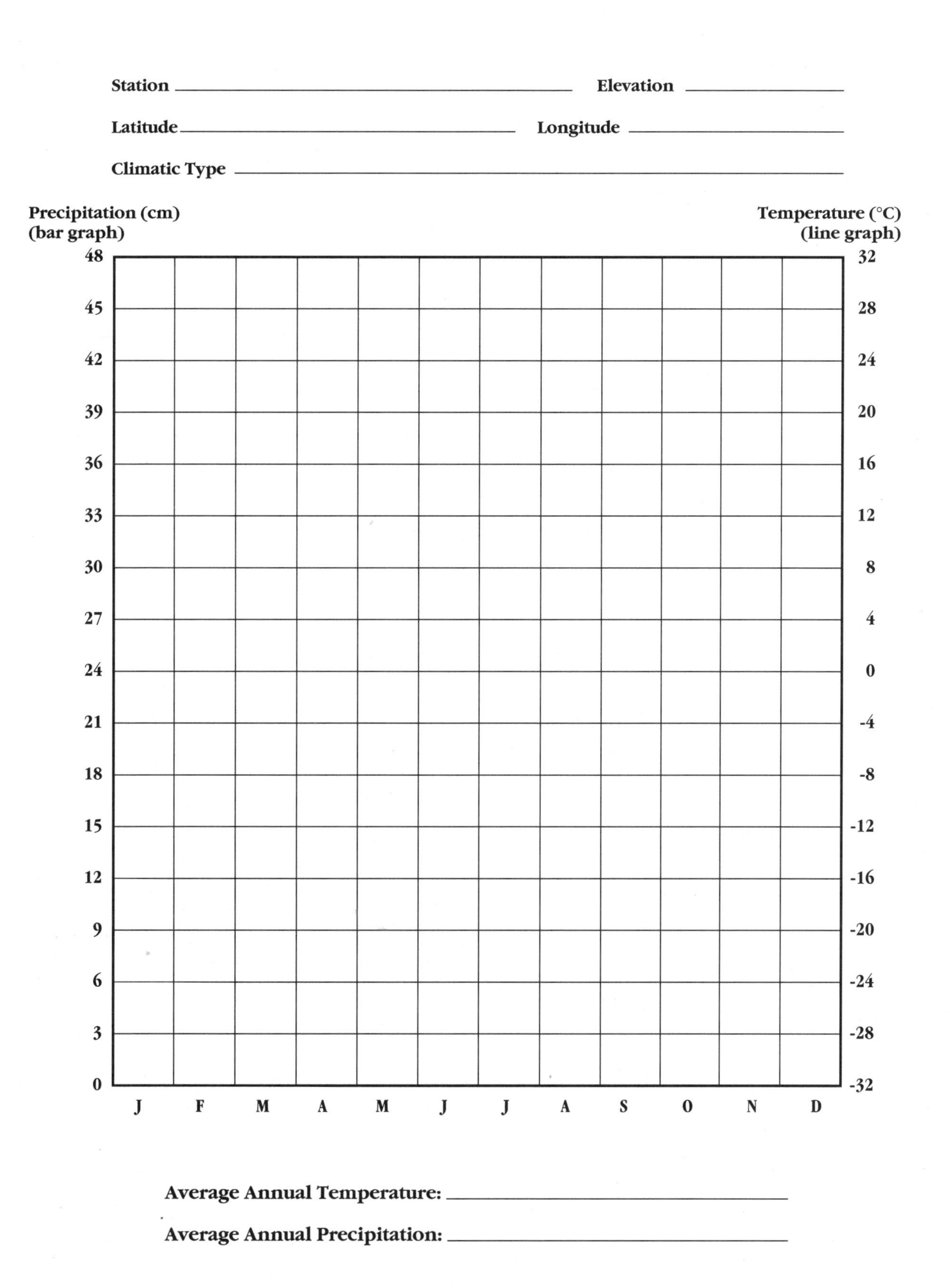
Station
Elevation
Latitude
Longitude
Climatic Type
Precipitation (cm)
(bar graph)
Temperature (°C)
(line graph)
48
45
42
39
36
33
30
27
24
21
18
15
12
9
6
3
0
32
28
24
20
16
12
8
4
0
-4
-8
-12
-16
-20
-24
-28
-32
J F M A M J J A S O N D
Average Annual Temperature:
Average Annual Precipitation:

TEACHER

PRESENTATION

2.1

Case Study of the Spoon River Watershed

Purpose

The purpose of this case study is to model the study of watersheds and location in order to prepare students for Student Activity 2.3: Investigating the Location of Watersheds.

Materials

For the teacher

transparencies of Spoon River Watershed; Estimation of Spoon River Watershed Area; Discharge of Spoon River at London Mills; and Climograph for Peoria, Illinois (Teacher Figures 2-3 through 2-6)
overhead projector
wall map of Illinois or of the United States
washable markers, several colors
ruler

Instructional Procedure

1. Explain that you will present a case study in which you will model the concepts and skills students will use in their investigation of watersheds and location in an upcoming activity (Student Activity 2.3).
2. Show the transparency of the Spoon River Watershed. Place an "X" at London Mills with a washable marker. Point out the location of London Mills on a wall map of Illinois or the United States.
3. Ask students to distinguish between relative location and absolute location. Describe the relative location of London Mills in any or all of the following ways: London Mills is in West Central Illinois, 35 miles west of Peoria, 160 miles southwest of Chicago, and on the Spoon River. Explain that the absolute location of London Mills is 42.27°N, 90.16°W, or round the numbers to 42°N, 90°W.
4. Use a washable marker to trace the river and its tributaries. Use another colored marker to draw the approximate watershed boundary. Ask students to explain the terms *watershed, tributaries,* and *divide.* Use these terms as you discuss the Spoon River watershed. Note that the Spoon River is itself a tributary of the Illinois River, which is in turn a tributary of the Mississippi River. Point out other watersheds that exist beyond the Spoon River divide: Kickapoo Creek and the Illinois River to the east, Mud Creek to the north, South Edwards River to the northwest, and so on.
5. Mark the mouth of the Spoon River with an "X." Ask students to describe the general direction from London Mills to the mouth of the Spoon River *(SSE).* Describe other directions, such as: Duncan is approximately

32 miles NE of London Mills, Roseville is W of London Mills and NW of the river's mouth, and so on.

6. Show the transparency of Estimation of Spoon River Watershed Area. Tell students that the map scale is 1:500,000 and 1 cm = 5 km. Explain the technique for estimating area by drawing squares and rectangles that approximate the shape of the watershed area. Measure and calculate the area of the two rectangles on the transparency:
   20 cm × 7.5 cm = 100 km × 37.5 km = 3750 square kilometers
   5 cm × 8 cm = 25 km × 40 km = 1000 square kilometers
   Add the two areas together:
   3750 + 1000 = 4750 square kilometers
   The answer provides an approximation of the actual watershed area of 4,800 square kilometers (1,855 square miles).
7. Show the transparency of Discharge of Spoon River at London Mills, which shows the average monthly discharge (in cubic meters per second) at the London Mills gauging station from October 1943 through September 1991. Point out that the volume varies greatly during the course of a year, with peak discharges occurring during winter and spring.
8. Show the transparency of Climograph for Peoria, Illinois, which shows average monthly temperatures and precipitation for Peoria from 1961 through 1990. Point out the seasonal range in temperature and the monthly distribution of precipitation.
9. Superimpose the Climograph for Peoria, Illinois, over the Discharge of Spoon River at London Mills. Ask students to explain how monthly temperature and precipitation fluctuations influence river discharge. (Desired response: *Precipitation is necessary to maintain stream flow. High temperatures during the summer cause high transpiration from plants and evaporation from land and water surfaces. The highest monthly discharges occur during the spring when rainfall is high and temperatures are low. Lowest discharges occur in the late summer when high temperatures increase evapotranspiration rates.*)

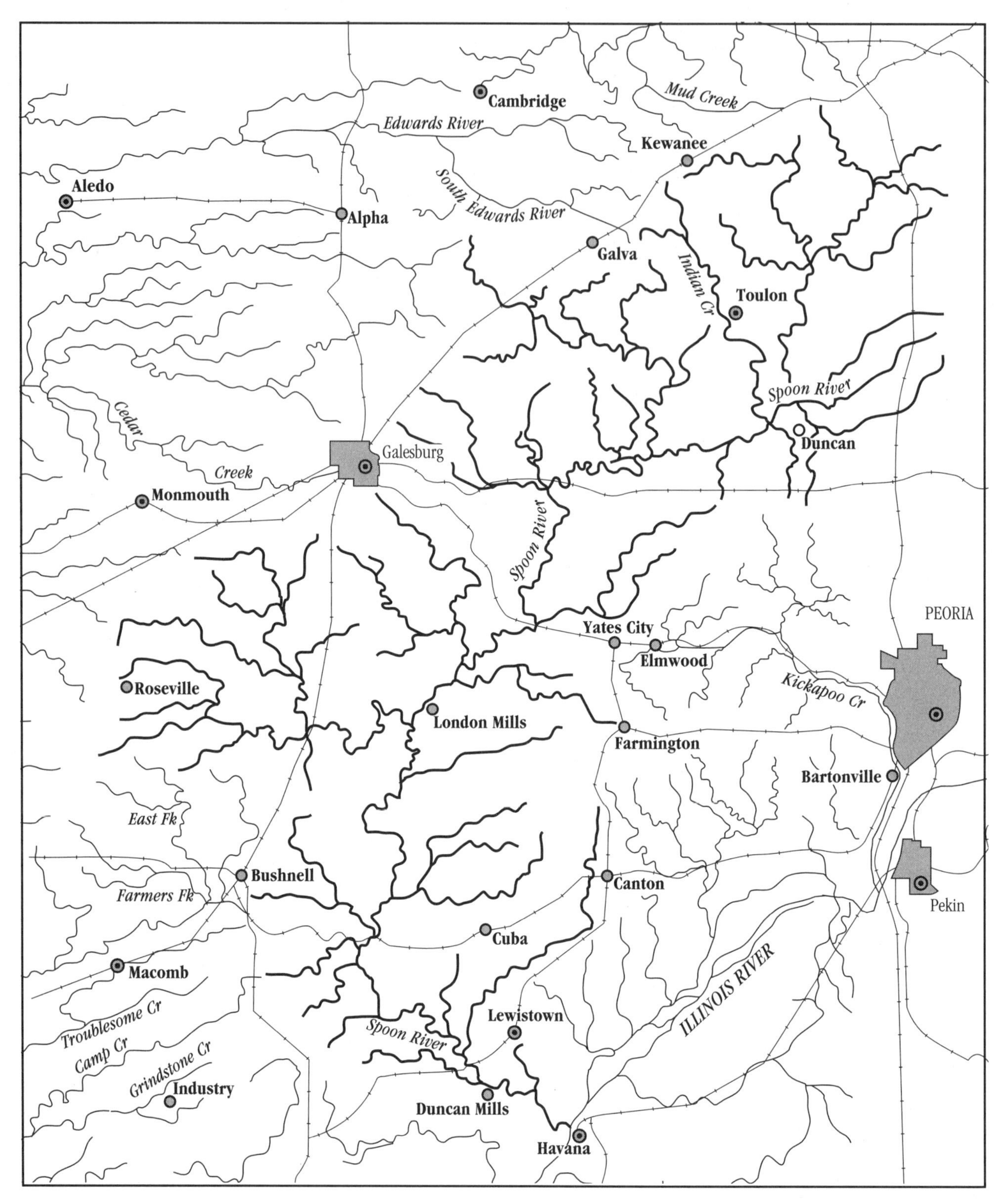

1 inch equals approximately 8 miles
1 cm = 5 km

Scale 1:500,000
Source: map of state of Illinois by U.S. Geological Survey

Teacher Figure 2-3: Spoon River Watershed

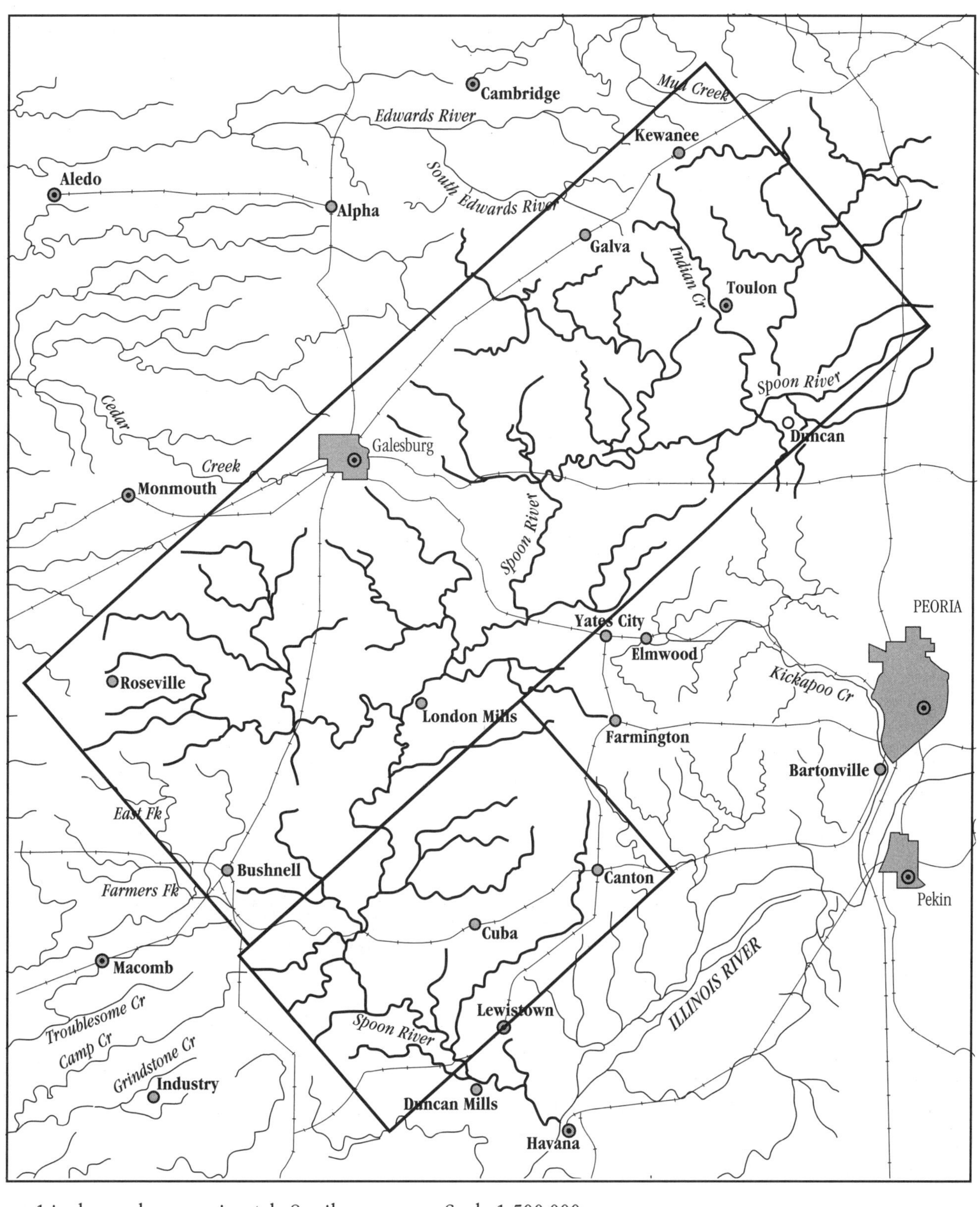

1 inch equals approximately 8 miles
1 cm = 5 km

Scale 1:500,000
Source: map of state of Illinois by U.S. Geological Survey

Teacher Figure 2-4: Estimation of Area of Spoon River Watershed

Discharge of Spoon River at London Mills (Mean monthly data, 1943–1991, in cubic meters/second)

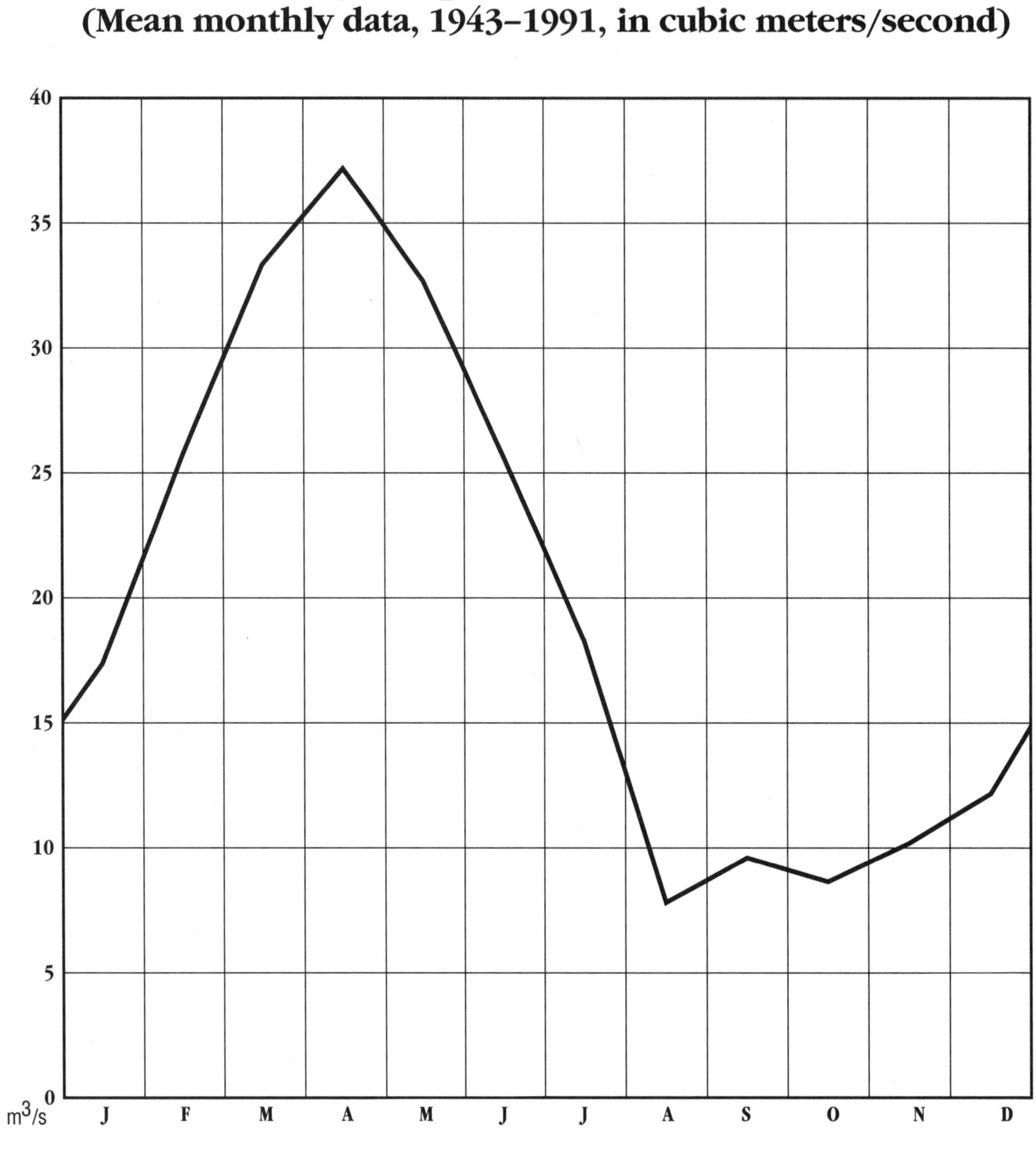

Source: Water Resources Data, Illinois Water Year 1991, Vol. 2, Illinois River Basin
U.S. Geological Survey Water-Data Report IL-91-2, pp. 329-330.
Data converted from discharge in cubic feet per second.

Teacher Figure 2-5: Discharge of Spoon River at London Mills

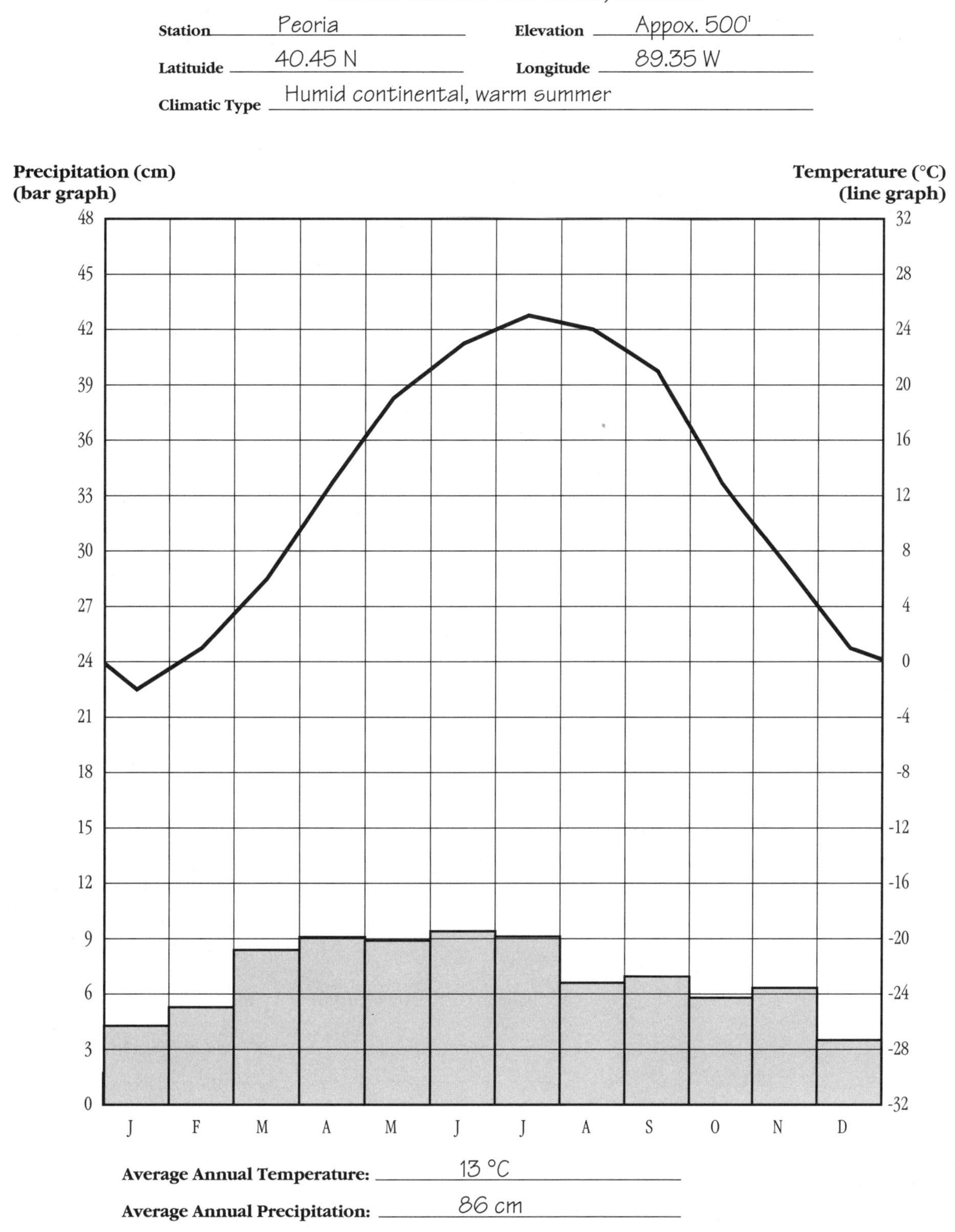

Teacher Figure 2-6: Climograph for Peoria, Illinois

T E A C H E R

For Student Activity
2.3

Investigating the Location of Watersheds

Purpose

In parallel with Student Activity 2.3: Investigating the Location of Watersheds, to provide specific teaching steps that facilitate teaching and classroom discussion of concepts relating to rivers and watersheds and discovery that various combinations of phenomena produce unique interrelationships at particular places. (Suggested answers are in italics; notes for teacher are in bold.)

Background

Students should read Student Information 2.1 and 2.2 before beginning Student Activity 2.3. **You may wish to assign a portion of this activity as homework.**

Students can answer Question 3 only if they have an atlas that has world temperature and precipitation maps. The following atlases contain such maps: *Goode's World Atlas, Nystrom Desk Atlas, Times Atlas of the World, New York Times Atlas of the World, World Book Atlas.* **(See Appendix C for full bibliographic citations.) For the materials list for this activity, see Student Activity 2.3 or the Teacher Notes for Lesson 2.**

Procedure

Ask students to recall the Spoon River case study while doing this activity.

1. Using an unlabeled map of North American rivers, place an "X" at the location of your school or community. Consult an atlas or other map to determine its relative and absolute location (latitude and longitude). Where is it located?

   Relative location (Use directions—N, S, E, W—where appropriate)

   Absolute location (to the nearest whole degree): latitude ____________

   longitude ____________

   Many answers are acceptable for identifying relative location; the school's location may be relative to other cities, the state, the region, or to physical features, such as the river or landforms.

   Point out students' responses to relative location of their school or community on a wall map or have students point them out. If students are having difficulty determining absolute location, refer them to Student Information 2.2. Insist that students use directions (N, S, E, W) where appropriate.

2. On your map, label in pencil the rivers that you can name.
   Answers will vary according to the five rivers that students select.

3. Look in your atlas and label other major rivers on your map. For five of the rivers you have labeled, determine the relative and absolute locations of the mouth of each river and determine the distance and direction of that river mouth from your school. Use a table like Data Table 1 shown on page 55.

DATA TABLE 1

Location of the Mouths of Five Rivers

Name of River	**Relative Location**	**Absolute Location**	**Distance from School (km/miles)**	**Direction from School**

Answers will vary according to the five rivers that students select.
Ask students to name the rivers they know before they refer to an atlas. Have volunteers point out the rivers they know on a wall map. Select a river and demonstrate the procedure for finding and reporting the information requested in the activity. If students still are having difficulty, refer them to Student Information 2.2.

4. Look up the terms *watershed, tributary,* and *divide* in the glossary and write a brief description of each in your own words.
   a. watershed *(the land area from which water drains toward a common watercourse in a natural basin.)*
   b. tributary *(a river or stream that contributes its water to a larger stream or river.)*
   c. divide *(a ridge or relatively high ground that separates two drainage basins.)*

   Draw on the chalkboard a hypothetical watershed showing a river, its tributaries, and a divide. Ask for questions before proceeding.

5. On your map of North American rivers, draw with a pencil the boundary of the Mississippi River watershed and the watershed of your local river. Then highlight the boundaries with a colored marker. Using an atlas, note the location of your local river watershed relative to the following: (Be sure to include directions—N, S, E, W—in all your answers.)
   Answers will depend on where the project watershed is. For answers to be acceptable, they should include directions, as shown in the following sample answers:
   a. Mississippi watershed *(Example: The Columbia River is W and NW of the Mississippi watershed.)*

b. state(s) in which your watershed is located *(Example: The Hiwassee River is located in N Georgia, SW North Carolina, and SE Tennessee.)*
c. North America *(Example: The Sacramento River watershed is located in Western United States in Northern California.)*
d. world *(Example: The Brazos River is located in the N and W hemispheres.)*
e. Is your watershed part of a larger watershed? If so, which one? *(Example: The Shenandoah River watershed is located in the southern part of the Potomac River watershed.)*

In order to make comparisons, students must draw boundaries for at least the Mississippi watershed and the project watershed. You may display a model of the map of North American rivers, marked so students can use it for comparing and correcting their own maps.

Assign additional watersheds as time and interest allow. For instance, you can facilitate discussion of the hierarchical arrangement of watersheds by having students draw boundaries of a watershed within the Mississippi watershed. Exact divides cannot be interpreted from the map and are not necessary for this exercise.

Observations

1. Using the map of your project watershed area, complete the following tasks:
   a. Highlight (with a highlighter or marker) the course of the river or stream and its tributaries.
   b. Plot the location of your school with a star (*).
   c. Draw the approximate boundary of the watershed with a pencil.
   d. With a different colored marker, highlight the watershed boundary.
   Answers will vary with watershed.

Analyses and Conclusions

1. Based on your labeled project watershed map, answer the following:
   You may wish to encourage the use of the terms *upper, middle,* and *lower* to express location relative to the total watershed. If you plotted latitude and longitude on the photocopied map, students can answer Question 2c without consulting other maps. If not, they will need to compare the large-scale photocopied map with a smaller scale map in an atlas. For large watersheds, the nearest whole or half degree of latitude should be acceptable. For smaller watersheds, you may instruct students to use the nearest minute of latitude and longitude if they are using large-scale maps.
   a. Where is your school relative to the whole watershed? *Answers will vary with watershed but should contain references to cardinal and intermediate directions.*
   b. Name the largest cities located within the watershed. *Answers will vary with watershed.*

c. If necessary, consult an atlas to find the absolute location of the watershed, and record to the nearest whole degree: *Answers will vary with watershed. Note that latitude is given first, followed by longitude.*
latitudinal range ________ (northernmost) to ________ (southernmost)
longitudinal range ________ (easternmost) to ________ (westernmost)

d. Calculate the approximate area of your watershed. (HINT: Use the map scale. Convert miles to kilometers by multiplying by 0.62. To determine the average length and width, use a ruler and draw two or three squares or rectangles approximating the watershed boundaries. Then measure the length and width of each and multiply to determine the area. Add the areas of the several squares or rectangles to arrive at the approximate area of your watershed. To convert square kilometers to square miles, multiply by 0.3861.) Show all your calculations.
Area of watershed = __________ square km __________ square miles
Answers will vary with watershed, but determine an acceptable range. Accept approximations, because minor measurement differences are exaggerated when multiplied.

If students are having trouble calculating watershed area, review the method for estimating area, using the transparency of Estimation of Spoon River Watershed Area as an example. Some students may require extra help with the mathematical procedures. Students may also need a refresher session on converting map scale (see Student Information 2.2). When students are finished, review by asking students to explain the procedures they used. If answers do not fall within an acceptable range, lead the class in calculating the area again. For more specific teaching activities on area and on converting map scale, see *Rivers Mathematics*.

2. If you have an atlas that has temperature and precipitation maps of the world, use that atlas to find a watershed of about the same size and approximately the same latitude on another continent. Consult the world temperature and precipitation maps, and a map with a relief legend, showing the elevation of the land above sea level. Then provide the following information:
   a. Name of the river
   b. Relative location within a country
   c. Names of major cities within the watershed
   d. Other than size, what similarities to the project river can you identify?
   e. What differences do you see between the watershed you've selected and the project watershed?
   f. Which of the two waterways would you expect to have the greater volume of discharge at its mouth? Why?

   Students can answer this question only if they have an atlas that has world temperature and precipitation maps. Answers will vary, but similarities and differences will probably focus on the following: elevation or relief, shape, length of river, amount and variability of precipitation, temperature range and variability of temperature, whether or not the watershed is tributary to another watershed, population (as gauged by the number and sizes of cities), transportation facilities, swamplands, marshes, and deltas.

Give students time to browse through the atlas to discover answers. Alert them to differences in the scales of the maps they are comparing. Because atlas maps will have a smaller scale and provide minimal detail, an exact match of watershed area cannot be expected. Point out the relief scale; you may wish to demonstrate how this scale applies to the project river watershed.

Critical Thinking Questions

For the following questions, you may consult the observations you have just made, Student Information 2.1 and 2.2, maps you have received, and the atlas.

1. Based on your observations, identify and describe the influence of three factors on river volume.
   Assuming similar size of watersheds, answers should cite precipitation, evapotranspiration, and temperature as variables that influence the volume of discharge.
2. Beyond the three factors that you have just listed, what other factors might influence the volume of a river's discharge?
   Other factors that may influence the volume of a river's discharge include seasonal variability of temperature and precipitation; withdrawal of water for irrigation and domestic or industrial use; reservoirs; amount and type of vegetative cover; and type of land surface.
3. Make a table in which you list five of the world's rivers that you would expect to have the greatest discharges at their mouths. Give the relative and absolute location of the mouth of each river.

Answers will vary but will probably often include the Amazon, Zaire (Congo), Mississippi-Missouri, and Ganges rivers .

After students have completed Critical Thinking Question 3, show the transparency of Largest Rivers by Discharge Volumes at their Mouths. Ask students to compare and discuss their selections.

As appropriate, have students complete the Keeping Your Journal section.

Largest World Rivers by Discharge Volumes at the Mouths (in cubic meters per second)

Lena
16,000 m^3/s

Yenisey
18,000 m^3/s

Mississippi
18,000 m^3/s

Madeira
22,000 m^3/s

Chang Jiang (Yangtze)
29,000 m^3/s

Rio Negro
30,000 m^3/s

Ganges-Bramaputra
31,000 m^3/s

Orinoco
36,000 m^3/s

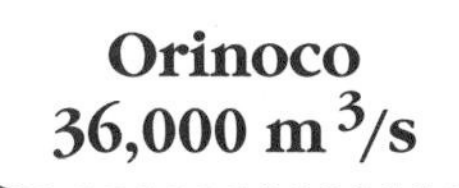

Zaire (Congo)
44,000 m^3/s

Amazon
200,000 m^3/s

Teacher Figure 2-7: Largest World Rivers By Discharge Volumes at Their Mouths (in cubic meters per second)

Name

River Systems

No matter where you live, a river or stream is probably not too far away. But have you ever thought about what a river is and what geographical terms you need to accurately describe a river? This information sheet will give you an overview of rivers and the river systems of which they are a part.

What Is a River?

A **river** is a body of water that flows into a larger body of water. A river consists of fresh water flowing in a **channel,** a path whose boundaries are marked by sloping ground called **banks.** Its bottom is called a **bed.** The beginning of a river or a **stream** (a relatively smaller body of running water) is its **source.** The source may be a spring, a lake, or a number of rivulets, or small streams, that join to form a bigger stream. (A **spring** is a continuous natural flow of water from the ground, occurring at the point where the water table intersects the land surface.) As smaller streams empty into a river, its path, or **course,** enlarges until it enters an ocean or other large body of water. The **mouth** is the point at which a stream or river flows into a larger body of water.

A **tributary** (TRIB yuh tehr ee) is a stream or river that contributes its water to a larger stream or river. The network of tributaries and the river into which they flow make up a **river system.** The area drained by a single river system is called a **drainage basin** or **watershed.** A drainage basin is separated from other basins by higher ground called a **divide.** The divide determines the basin into which precipitation will flow. A major drainage basin may contain smaller contributing systems, which in turn may contain still smaller systems.

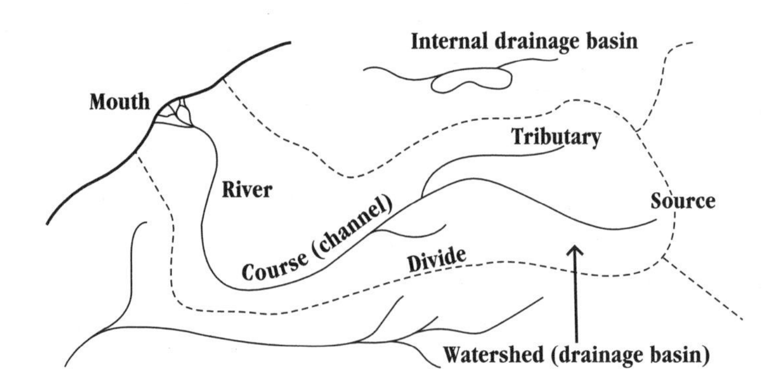

Figure 2-1: The Area Within a Divide is a Drainage Basin or Watershed

How Does Precipitation Affect a River?

Rivers are influenced by the amount, frequency, and type of precipitation that falls in their drainage basin. In regions of heavy precipitation, rivers carry a greater volume (cubic meters) of water and have a higher total **discharge** (cubic meters per second) flowing into a larger body of water.

The most common type of precipitation is rain. Following a heavy rainfall, water pools on the surface of the earth. Some of the pooled water sinks into the earth, replenishing groundwater supplies. Surface water accumulates in streams, which merge into rivers and carry the runoff to larger bodies of water at lower elevations. Eventually, most surface water reaches the oceans. Some drainage basins are not connected to the oceans. Such systems, called **internal drainage** systems, are found in dry areas where river volume is insufficient to provide continuous flow beyond the area where the water pools. High evaporation rates return water to the atmosphere; some internal "lakes" may be dry part of the year. The Great Salt Lake in Utah, the Caspian and Aral Seas in Asia, Lake Chad in Africa, and Lake Eyre in Australia are examples of **saline lakes,** lakes in internal drainage systems.

A look at several of the world's rivers may help you see how precipitation and other factors influence rivers and their surrounding areas. Brief descriptions of the Amazon in South America, the Nile in Africa, and the Ganges-Bramaputra in South Asia follow.

Driftwood

"He once asked him, 'Have you also learned that secret from the river; that there is no such thing as time?'

A bright smile spread over Vasudeva's face.

"'Yes, Siddhartha,' he said. 'Is this what you mean? That the river is everywhere at the same time, at the source and at the mouth, at the waterfall, at the ferry, at the current, in the ocean and in the mountains, everywhere, and that the present only exists for it, not the shadow of the past, nor the shadow of the future?'"

From *Siddhartha* by Herman Hesse

The Amazon River

The Amazon River basin is the world's largest, covering 5,800,000 square kilometers (2,262,000 square miles) of South America, mainly in Brazil. Combined with the Ucayali, which itself is nearly 2,000 kilometers (1,240 miles) long, the Amazon is approximately 6,450 kilometers (4,000 miles) long. Several other tributaries of the Amazon—including the Juruá Madeira, Marañón Negro, Purus, and Xingu—are more than 1,600 kilometers (992 miles) long, ranking them among the world's longest. This vast watershed sits astride the equator in a region of very high precipitation. The enormous size of the watershed and the high precipitation in the region combine to give the Amazon the greatest volume and discharge of all rivers.

The Nile River

The Nile is the longest river in the world, stretching more than 6,600 kilometers (4,100 miles) north from its source. The Nile begins with small streams that feed Lake Victoria in Africa's Eastern Highlands. The mouth of the Nile is on the coast of the Mediterranean Sea. The Nile drains an area of approximately 3,000,000 square kilometers (1,238,000 square miles), which is about half the size of the Amazon drainage basin. More than half the Nile's length stretches through the extremely dry desert regions of North Africa. A river such as the Nile that flows through a desert is called an **exotic river.**

In the northern desert region, the Nile loses more water through evaporation than it gains from precipitation, so the discharge of the Nile at its mouth is much less than the discharge of the Amazon. In the hot, dry climate, water is rapidly lost through the very high rate of **evapotranspiration** (ih VAP oh tran spuh RAY shuhn), which is the change of water into vapor by evaporation from land and water surfaces and by transpiration from plants. Because of the extreme dryness of the air masses over North Africa, very little of the water vapor condenses and falls again as rain in the region. Most of Egypt's approximately 60 million people live along the Nile, and they use water from the Nile for crop irrigation and domestic purposes, further decreasing the river's volume.

The Nile's Aswan High Dam was completed in 1971 in southern Egypt. Behind the dam lies Lake Nasser, one of the world's largest artificial **reservoirs,** a body of stored water having a capacity of 162,000,000 cubic meters. The dam produces **hydroelectricity** (hi droh ih lehk TRIHS uht ee) by converting the energy of moving water into electricity. The dam also controls flooding and provides water for irrigation. The continuous flow of irrigation water allows Egyptian farmers to grow crops year around.

Changing the nature of the Nile has not been without problems. Before the completion of the dam, annual flooding of the river deposited nutrients on the **floodplains**—the flat expanses of land that border a river and become covered with water when the river overflows. Without these nutrients, Egyptian farmers must apply costly artificial fertilizers to their lands. The dam has also reduced the amount of nutrients that flow into the Mediterranean, causing a decline in sardine populations and hurting the fishing industry. The high rate of evaporation has contributed to **salinization** (sal uh nuh ZAY shuhn), or the buildup of salts, on irrigated lands. In addition, some researchers have suggested that the slowed current of the Nile may be related to the increased incidence of diseases caused by river-dwelling parasites.

The Ganges and Bramaputra Rivers

The country of Bangladesh (bahn gluh DEHSH) occupies the massive **delta** at the mouth of the Ganges (GAHN jeez) and Bramaputra (brahm uh POO truh) Rivers. A delta is formed by the **deposition** (dehp uh ZISH uhn), or laying down, of **alluvium** (uh LOO vee uhm), consisting of silt, sand, and fine rock particles.

Bangladesh is frequently flooded because the elevation of the delta is not much above sea level. **Winds,** the movement of air caused by the uneven heating and cooling of earth's surface, have very powerful effects on climate

Driftwood

"The sea lies all about us. The commerce of all lands must cross it. The very winds that move over the lands have been cradles on its broad expanse and seek ever to return to it. The continents themselves dissolve and pass to the sea, in grain after grain of eroded land. So the rains that rose from it return again in rivers. In its mysterious past it encompasses all the dim origins of life and receives in the end, after, it may be, many transmutations, the dead husks of that same life. For all at last returns to the sea—to Oceanus, the ocean river, like the ever-flowing stream of time, the beginning and the end."

From *The Sea Around Us* by Rachel Carson

in many parts of the world. In the case of Bangladesh, moisture-bearing winds, called **wet monsoons,** bring heavy rainfall to the region from May through October. Flooding has been aggravated by the removal of trees and other plants from the slopes of the Himalayan Mountains and other mountain ranges that form a divide between the rivers.

DRIFTWOOD

**"The thirsty earth soaks up the rain,
And drinks, and gapes for drink again."**

From *Drinking* by Abraham Crowley

Questions

Write your answers on a separate piece of paper. Use complete sentences.

1. What is the difference between a river and a river system?
2. How do precipitation, temperature, and evaporation influence the character of a river?
3. Why does the Amazon have the greatest volume and discharge of all rivers?
4. What effects has the Aswan High Dam had on the Nile River?
5. Why is Bangladesh frequently flooded?

Map Scale, Direction, and Location

Everything is somewhere. Geographers answer the question "Where?" by describing the locations of people and other phenomena on Earth's surface. Geographers are interested in understanding how locations and regions on Earth are similar and different. They seek to find the **spatial** (SPAY shuhl) **distribution,** or **patterns** resulting from the way **phenomena** (fih NAHM uh nuh)—observable facts and events—are distributed across part or all of Earth's surface. To see these patterns, geographers simplify the real world on symbolic representations called **maps.**

Scale

Actual distances on Earth must be represented in a greatly reduced form to fit conveniently on a map. **Map scale** is the ratio of distance on a map to the actual distance that is represented. For example, distances on a 1:2 (one to two) scale map would be half of the actual distances on Earth.

A map drawn at a large scale can show a great amount of detail. For this reason, a map of your school campus might be drawn at a scale of 1:1,000 or 1:2,000. By contrast, a small scale of 1:20,000,000 or 1:30,000,000 might be selected to show the entire world on a wall map. A 1:3,000,000 scale map represents Earth's surface at one three millionth (1/3,000,000) its real size. The ratio of map area to actual surface area of the earth is called a **representative fraction.** A representative fraction of 1/3,000,000 means that one unit of measure represents 3,000,000 of the same units on Earth's surface.

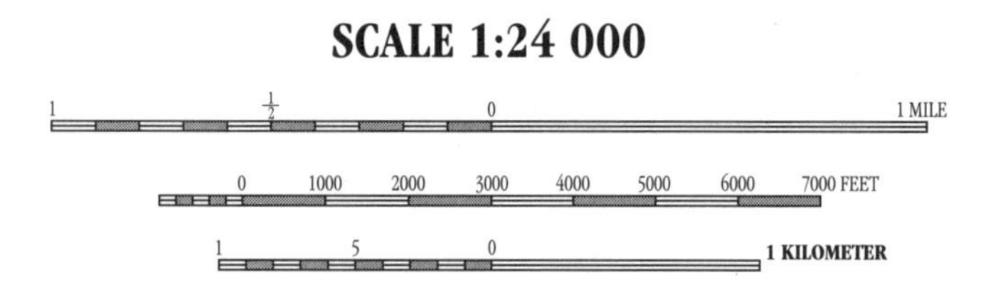

Source: U.S. Geological Survey

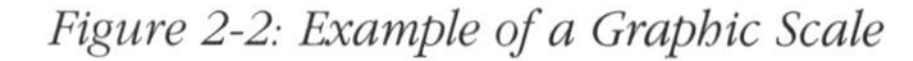

Figure 2-2: Example of a Graphic Scale

Some maps use a **stated scale,** which is an expression of the ratio of the distance on a map to the actual distance represented. For example, a stated scale for a 1:3,000,000 scale map might be 1 centimeter equals 30 kilometers (1 cm = 30 km), or 1 inch equals approximately 47 miles (3,000,000 divided by 63,360 inches in one mile). Maps also may use a **graphic scale,** which is a bar scale that can be used to measure distances.

Direction

Direction on a map is indicated either by an arrow that points to geographic north (the North Pole) or by a compass rose that shows the four **cardinal directions** of north, south, east, and west (N, S, E, W). The compass rose also may indicate more precise directions, called **intermediate directions,** such as NNE, NE, ENE, ESE, SE, SSE, SSW, SW, WSW, WNW, NW, and NNW. Together, cardinal and intermediate directions are called **compass directions.** Generally, but not always, maps are oriented with north at the top of the map.

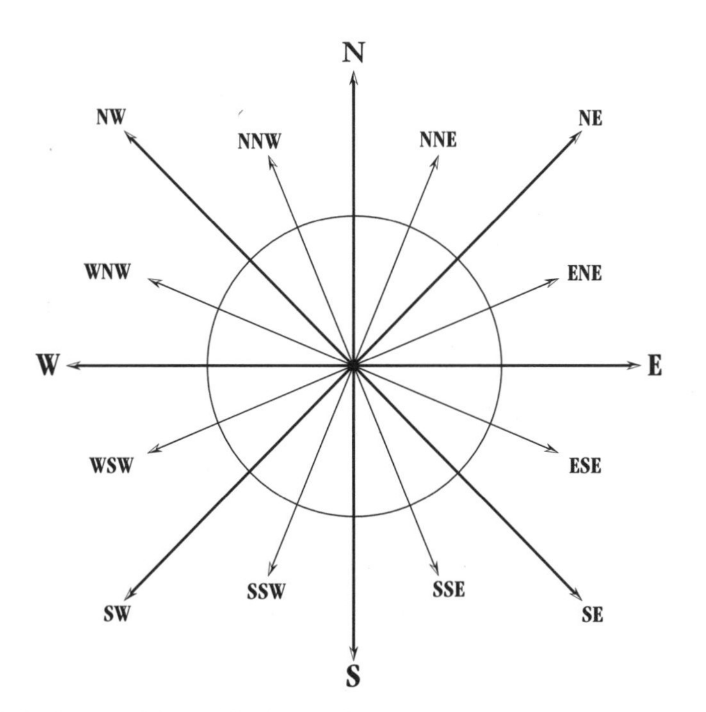

Figure 2-3: Compass Rose with Cardinal and Intermediate Directions

Location

You will remember from the first lesson that location refers to a position on Earth's surface. Geographers use a **geographic grid,** or system of imaginary lines on a map, to indicate **absolute location**—a precise and unique position on Earth. Basic to this system is the position of the **equator,** an imaginary line that circles the earth exactly midway between the North Pole and the

South Pole. The equator divides the earth into two half-spheres, referred to as the northern and southern **hemispheres** (HEHM uh sfeers).

Absolute location is usually described by the use of a latitude-longitude grid or an alpha-numeric system. **Latitude** is the distance in degrees north or south of the equator. The equator is assigned a value of 0 degrees latitude. The North Pole is assigned a value of 90 degrees north. The South Pole is assigned a value of 90 degrees south. Latitudinal lines are called **parallels** because they circle the earth parallel to the equator. Parallels of latitude are used to measure the distance north or south of the equator.

Longitudinal lines, called **meridians** (muh RID ee uhns), run from pole to pole. **Longitude** (LAHN juh tyood) is the distance in degrees east or west of the **prime meridian,** which is assigned the value of 0 degrees longitude. Because longitudinal lines converge at the poles, they are not parallel. The prime meridian, 0 degrees longitude, and its opposite, 180 degrees longitude, divide the earth longitudinally into eastern and western hemispheres.

When the two systems of lines are superimposed, they form a latitude-longitude grid. Absolute locations can be stated by plotting the distance north or south of the equator and the distance east or west of the prime meridian. For additional precision, each degree can be subdivided into 60 **minutes** of latitude or longitude, and each minute can be divided into 60 **seconds.** The Global Positioning System (GPS) uses satellite communications technology to establish absolute location.

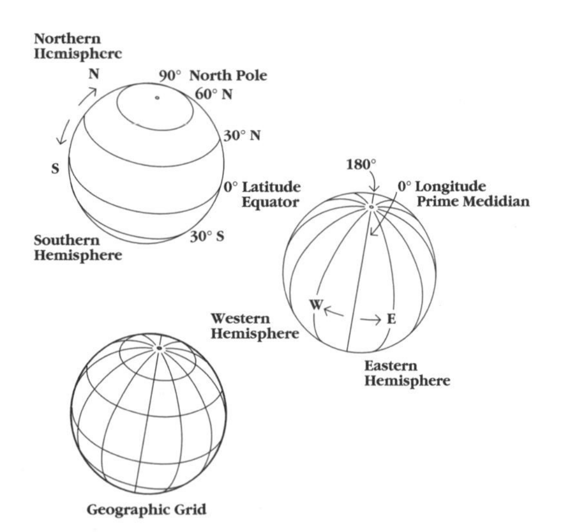

Figure 2-4: Latitudinal and Longitudinal Lines

Absolute location can also be shown on maps by using an alpha-numeric grid. The location of the desired item is described as the area where the alphabetic and the numeric grid rows and columns intersect. State highway maps commonly use the alpha-numeric system.

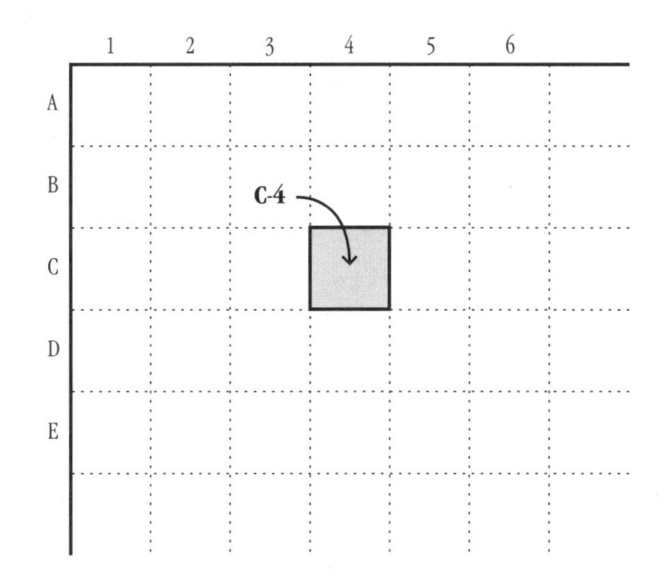

Figure 2-5: Alpha-Numeric System

Geographers are not satisfied with knowing only the absolute locations of phenomena. They also want to know the relationship of one feature or system to others. For example, a large city may have grown up along a river to take advantage of the river as a transportation system or as a source of power. The construction of a new bridge across the river may encourage new activities because the two sides, previously separated by the river, now are more accessible to one another. The "Mississippi Flyway" is important to waterfowl migration because of the large amount of wetland associated with the river. These examples illustrate relative location and help the geographer to understand another important question: "Why is it located where it is?" Changing one system may impact another, altering their relative locations.

Relative location is the position on Earth that is indicated with reference to the locations of other phenomena. Relative location can be used to show interrelationships of several phenomena. Because such relationships vary, each location is unique. For example, the Iranian city of Ābādān (ahb uh DAHN) is located at 30.20 N, 48.16 E. Using absolute location, you can find the city on a map at the point where the coordinates intersect: 30 degrees, 20 minutes north of the equator and 48 degrees, 16 minutes east of the prime meridian. In order to understand what makes Ābādān unique, however, you must realize the importance of its location relative to other phenomena. For instance:

1. Ābādān is located in the Iranian province of Khuzistan. Khuzistan's SW boundary is also Iran's national boundary with neighboring Iraq.
2. Ābādān is located on the delta—a swampy lowland at the head of the Persian Gulf—formed by the Shatt al Arab (shat al AR uhb) and south-flowing Kārūn rivers.

DRIFTWOOD

"When I'm playful I use the meridians of longitude and parallels of latitude for a seine, and drag the Atlantic Ocean for whales!"

From *Life on the Mississippi* by Mark Twain

3. Ābādān is on the east bank of the Shatt al Arab River on Ābādān Island. The city is approximately 50 km (31 miles) from the river's mouth on the Persian Gulf.
4. The Tigris and Euphrates rivers are tributaries that merge to form the Shatt al Arab, which flows approximately 200 km (124 miles) in a generally SE direction to the Persian Gulf.
5. Ābādān is located in a vast petroleum-producing region. Oil companies constructed not only pipelines to connect the city to the refinery, but also shipping terminals.
6. Just upstream from Ābādān are two cities that serve much the same purpose as Ābādān : the Iranian city of Khorramshahr and the Iraqi city of Al Basrah or Basra (BAHZ ruh).
7. The Shatt al Arab serves as part of the political boundary between Iran and Iraq. Because Iraq's international boundary on the Persian Gulf is squeezed to only a few kilometers between Iran on the north and Kuwait on the south, the river is critical to international trade. Iran and Iraq fought a war from 1980 to 1988, partly over control of the waterway. Ābādān's port facilities were largely destroyed during the conflict.

As geographers digest facts such as those described here, they attempt to develop pictures in their minds, or construct mental maps, to understand spatial relationships.

Questions

Write your answers on a separate piece of paper. Use complete sentences (except for the sketch).

1. Review all items in this handout that describe Ābādān 's relative location; then sketch a map that illustrates the relative locations of the place names mentioned. Use a pencil. Then compare your sketch map with a map in an atlas or provided by your teacher.
2. What does a map scale of 1:3,000,000 mean?
3. What is the difference between cardinal directions and intermediate directions?
4. Give the relative location and the absolute location of Ābādān.

Name

Investigating the Location of Watersheds

Purpose

To practice concepts relating to rivers and watersheds and to discover that various combinations of phenomena produce unique interrelationships at particular places.

Background

Make sure you have read Student Information 2.1 and 2.2 before beginning this activity. Consult available maps as necessary and be prepared to defend the information you gather.

Procedure

1. Using the unlabeled map of North American rivers (Figure 2-6), place an "X" at the location of your school or community. Consult an atlas or other map to determine its relative and absolute location (latitude and longitude). Where is it located?
   Relative location (Use directions—N, S, E, W—where appropriate) __________
   Absolute location (to the nearest whole degree) :
   latitude __________
   longitude __________
2. On your map, label in pencil the rivers that you can name.
3. Look in your atlas and label other major rivers on your map. For five of the rivers you have labeled, determine the relative and absolute locations of the mouth of each river and determine the distance and direction of that river mouth from your school. Use a table like Data Table 1 shown on page 55.
4. Look up the terms *watershed, tributary,* and *divide* in the glossary and write a brief description of each in your own words.
5. On your map of North American rivers, draw with a pencil the boundary of the Mississippi River watershed and the watershed of your local river. Then highlight the boundaries with a colored marker. Using an atlas, note the location of your local river watershed relative to the following. (Be sure to include directions—N, S, E, W—in all your answers.)
   a. Mississippi watershed
   b. state(s) in which your watershed is located
   c. North America
   d. world
   e. Is your watershed part of a larger watershed? If so, which one?

Materials

Per student

- photocopied portion of state highway map or state geological survey map showing project watershed, or state highway map
- atlas
- markers and highlighters
- ruler
- pencil
- calculator (optional)

Per class

- additional relevant atlases and maps (particularly atlases with world temperature and precipitation maps)

Observations

1. Using the map of your project watershed area, complete the following tasks:
   a. Highlight (with a highlighter or marker) the course of the river or stream and its tributaries.
   b. Plot the location of your school with a star (*).
   c. Draw the approximate boundary of the watershed with a pencil.
   d. With a different colored marker, highlight the watershed boundary.

Analyses and Conclusions

1. Based on your labeled project watershed map, answer the following:
   a. Where is your school relative to the whole watershed?
   b. Name the largest cities located within the watershed.
   c. If necessary, consult an atlas to find the absolute location of the watershed, and record to the nearest whole degree:
   latitudinal range _________ (northernmost) to _________ (southernmost)
   longitudinal range _________ (easternmost) to _________ (westernmost)
   d. Calculate the approximate area of your watershed. (HINT: Use the map scale. Convert miles to kilometers by multiplying by 0.62. To determine the average length and width, use a ruler and draw two or three squares or rectangles approximating the watershed boundaries. Then measure the length and width of each and multiply to determine the area. Add the areas of the several squares or rectangles to arrive at the approximate area of your watershed. To convert square kilometers to square miles, multiply by 0.3861.) Show all your calculations.
   Area of watershed = _________ square km _________ square miles
2. If you have an atlas that has temperature and precipitation maps of the world, use that atlas to find a watershed of about the same size and approximately the same latitude on another continent. Consult the world temperature and precipitation maps, and a map with a relief legend, showing the elevation of the land above sea level. Then provide the following information:
   a. Name of the river
   b. Relative location within a country
   c. Names of major cities within the watershed
   d. Other than size, what similarities to the project river can you identify?
   e. What differences do you see between the watershed you've selected and the project watershed?
   f. Which of the two waterways would you expect to have the greater volume of discharge at its mouth? Why?

DATA TABLE 1

Location of the Mouths of Five Rivers

Name of River	Relative Location	Absolute Location	Distance from School (km/miles)	Direction from School

Critical Thinking Questions

For the following questions, you may consult the observations you have just made, Student Information 2.1 and 2.2, maps you have received, and the atlas.

1. Based on your observations, identify and describe the influence of three factors on river volume.
2. Beyond the three factors that you have just listed, what other factors might influence the volume of a river's discharge?
3. Make a table in which you list five of the world's rivers that you would expect to have the greatest discharges at their mouths. Give the relative and absolute location of the mouth of each river.

Keeping Your Journal

1. Select one of the rivers about which you have found information for this activity and explain why you would like to visit it.
2. Have you ever waded in a mountain stream? Seen a large spring? Viewed a waterfall? Seen a river from an airplane? If so, describe the sensations you experienced.
3. Why did early settlers select a particular river site for your community? How does it compare with the locations of other communities?
4. Write a paragraph or two that completes the following: My community (park, highway, residential development, business) benefits from its location relative to a river because...

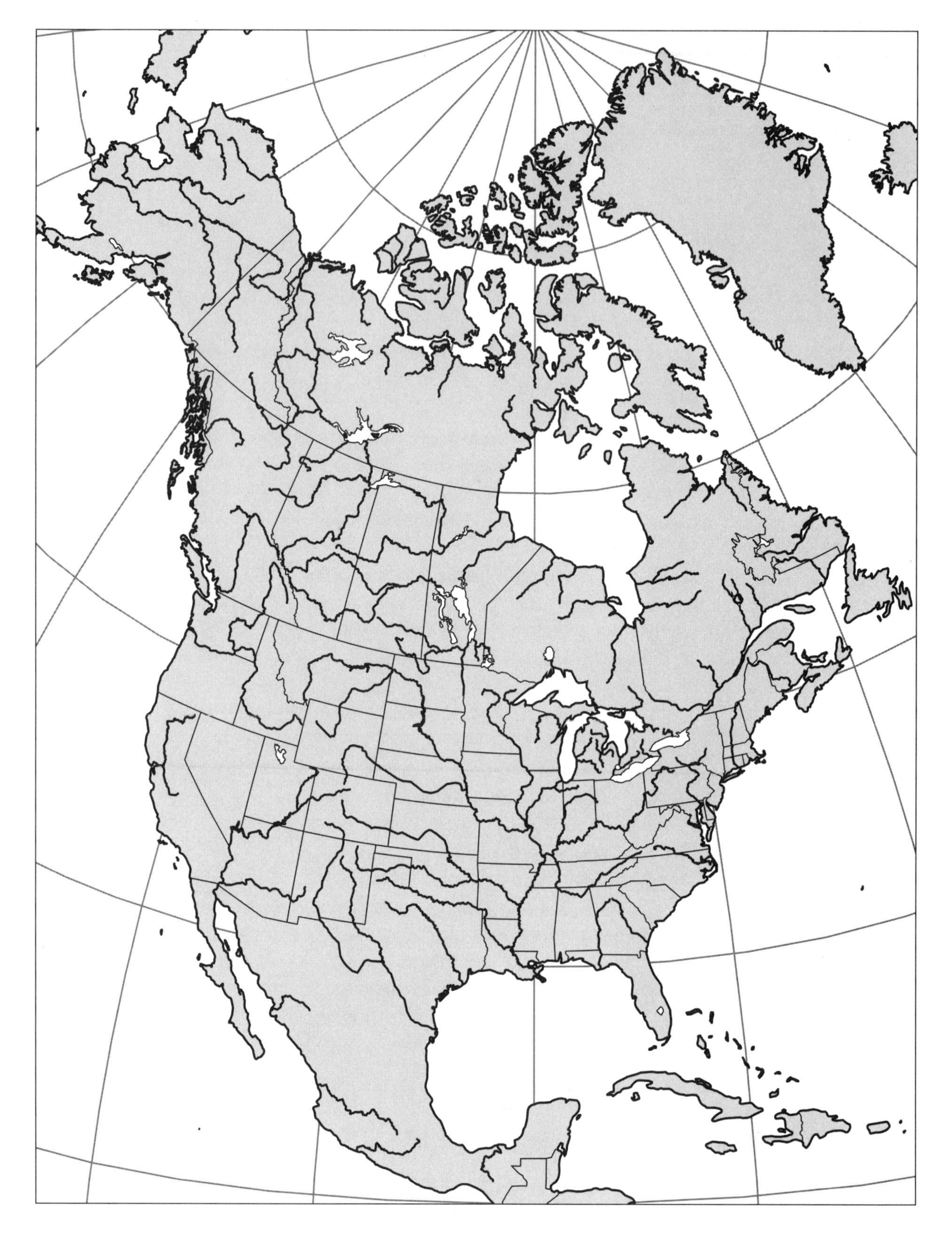

Figure 2-6: North American Rivers

Name

STUDENT ASSESSMENT 

Relating Location to Characteristics of Rivers

Introduction

In this assessment activity, you will match temperature and precipitation data with selected cities and the rivers on which they are located. For each city, the **data** (information in the form of facts or figures) appear in the form of a **climograph,** a graph that illustrates the temperature and precipitation averages for a particular location.

Except for Dawson, all the cities are located at the mouths of the rivers. For Buenos Aires—located on the Río de la Plata estuary near the mouth of the Río Paraná—only the discharge for Río Paraná is given. Using the information provided, you must justify every match you make by explaining the relationships between location and temperature, precipitation, and volume of river discharge.

Procedure

1. Study the location of each city and river on the world map, Cities on Major Rivers of the World. Make as many observations as you can about each location. Record your observations on a separate piece of paper in your own Data Table 1.
2. Examine carefully the eight temperature-precipitation graphs, or climographs. List each city and river in the first column of your own Data Table 2. In the second column, list the letter of the climograph that matches each city and river. You may use maps that show temperature and precipitation to help you do these matches.

Analyses and Conclusions

Answer this question on a separate piece of paper.

1. For each city and river, write several sentences explaining why you related a particular graph to that city and river.

Observations

DATA TABLE 1

City/River (list each city)	Observations

DATA TABLE 2

City/River (list each city)	Climograph (station letter)

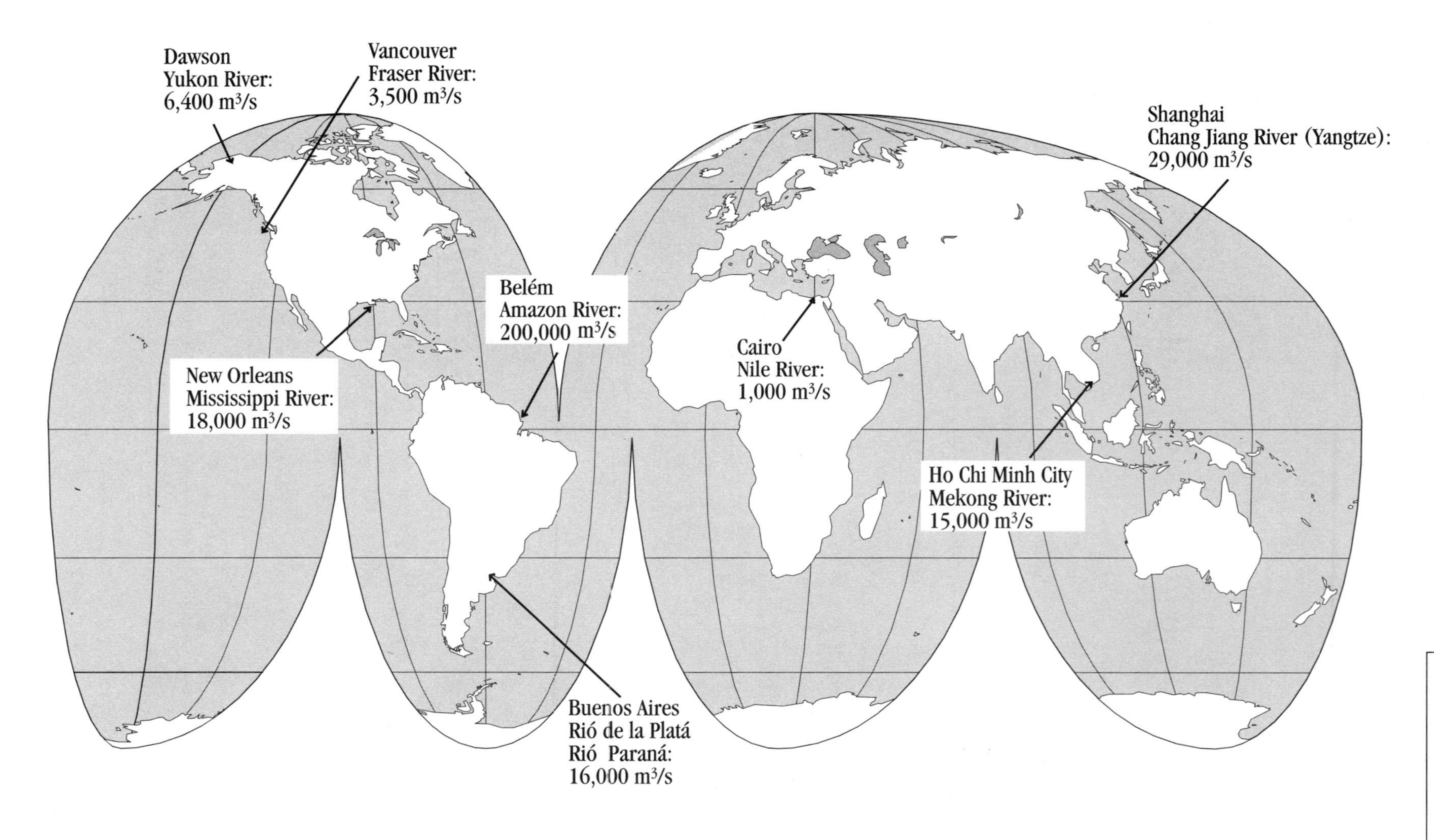

Figure 2-7: Cities on Major Rivers of the World

CLIMOGRAPH for Station A

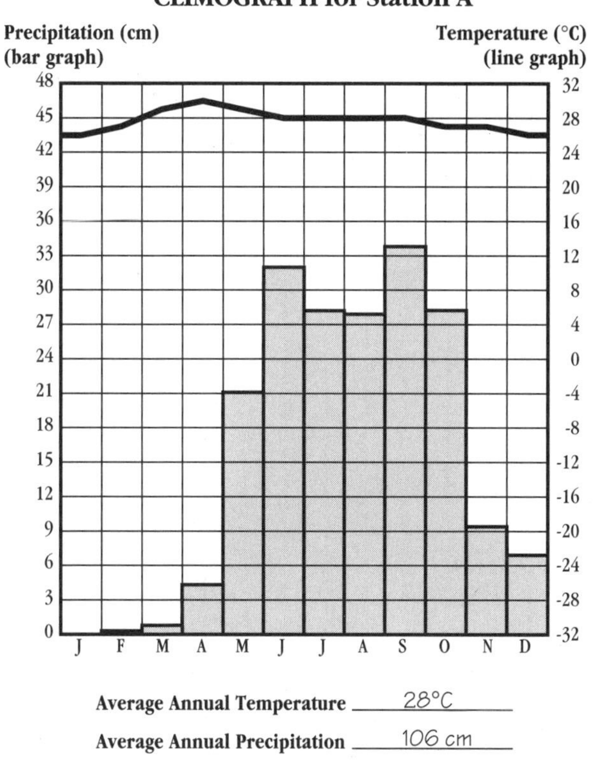

Average Annual Temperature 28°C

Average Annual Precipitation 106 cm

CLIMOGRAPH for Station B

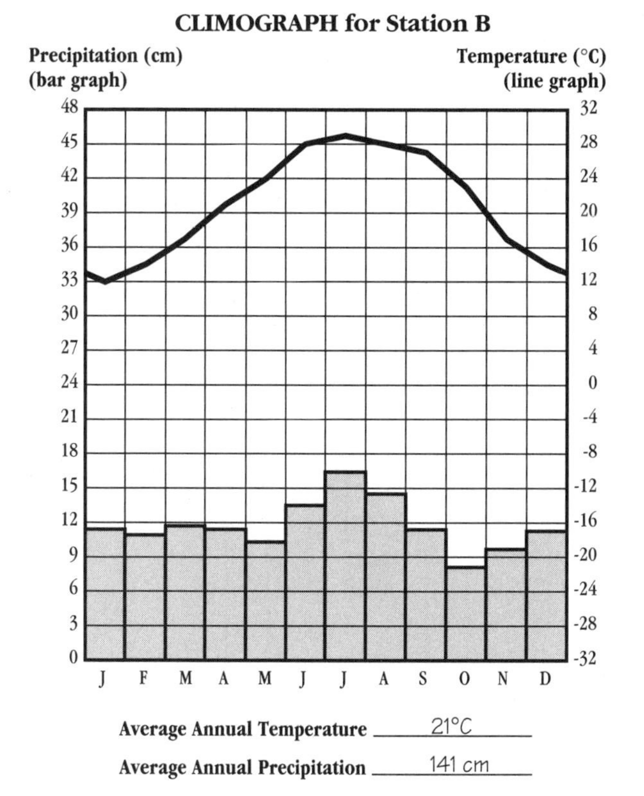

Average Annual Temperature 21°C

Average Annual Precipitation 141 cm

CLIMOGRAPH for Station C

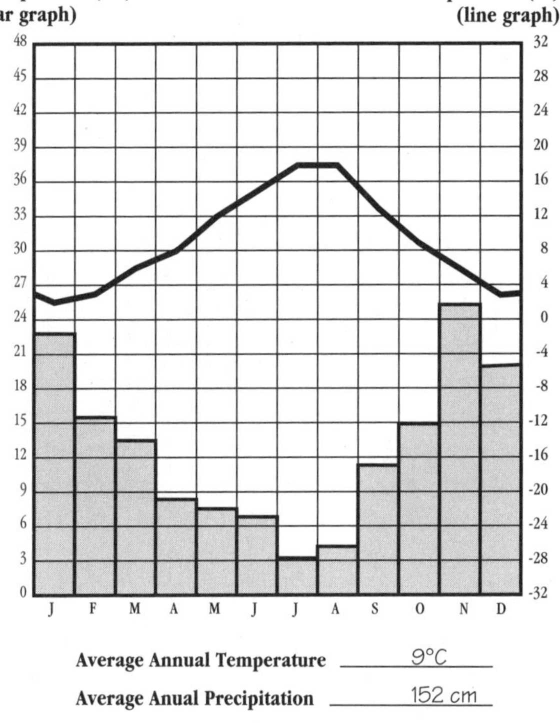

Average Annual Temperature 9°C

Average Anual Precipitation 152 cm

CLIMOGRAPH for Station D

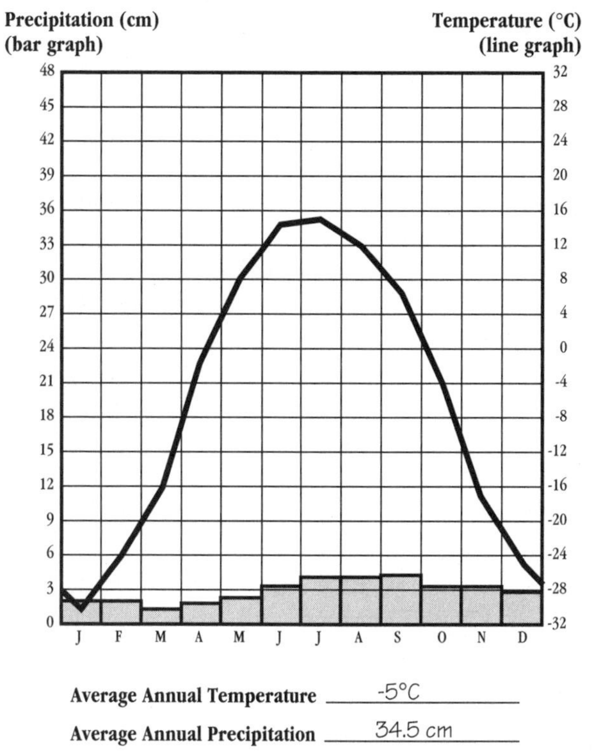

Average Annual Temperature -5°C

Average Annual Precipitation 34.5 cm

Climographs for Stations A through D

CLIMOGRAPH for Station E

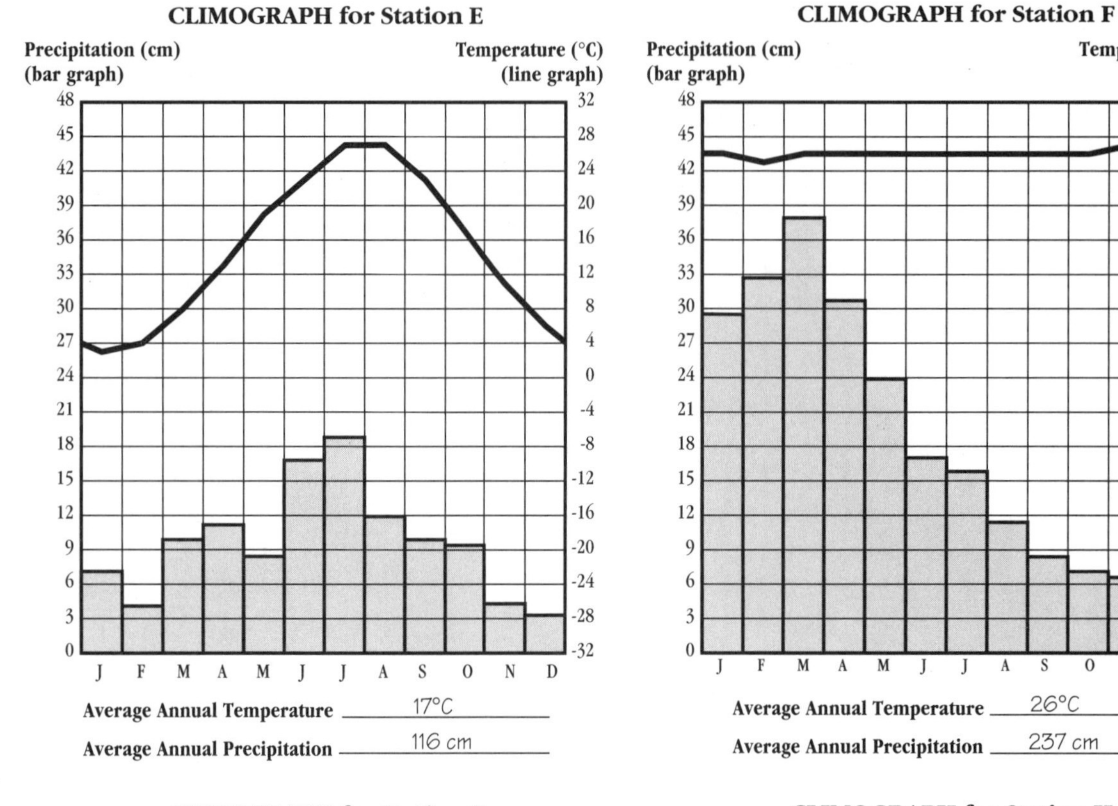

Average Annual Temperature 17°C

Average Annual Precipitation 116 cm

CLIMOGRAPH for Station F

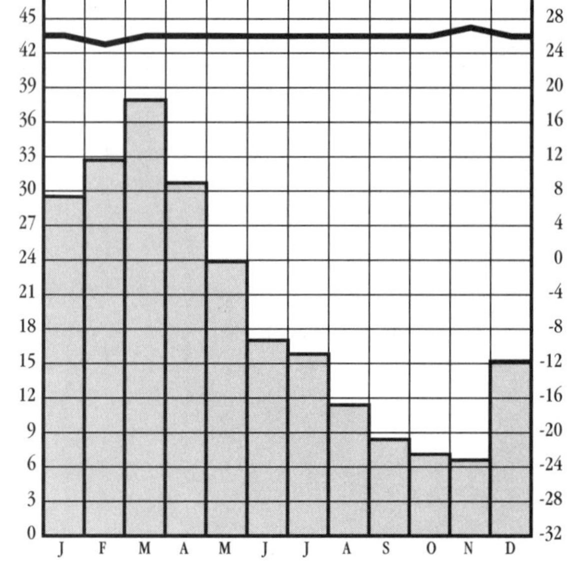

Average Annual Temperature 26°C

Average Annual Precipitation 237 cm

CLIMOGRAPH for Station G

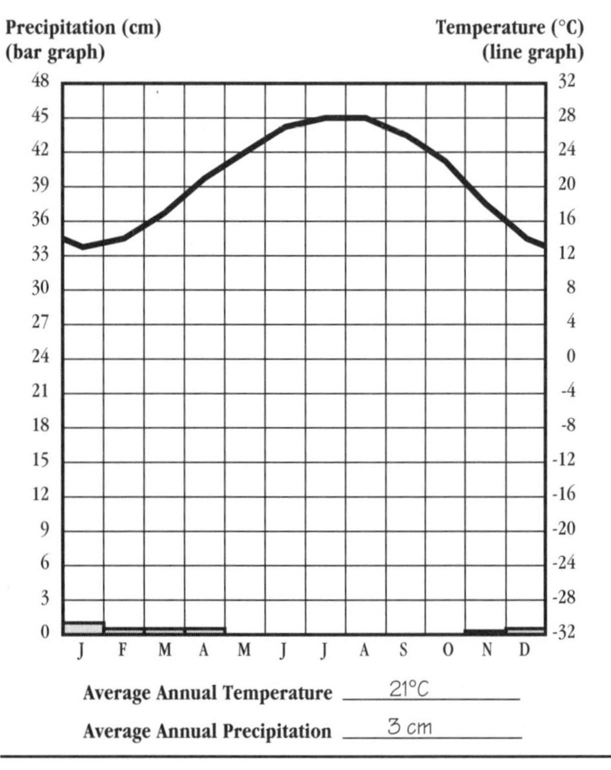

Average Annual Temperature 21°C

Average Annual Precipitation 3 cm

CLIMOGRAPH for Station H

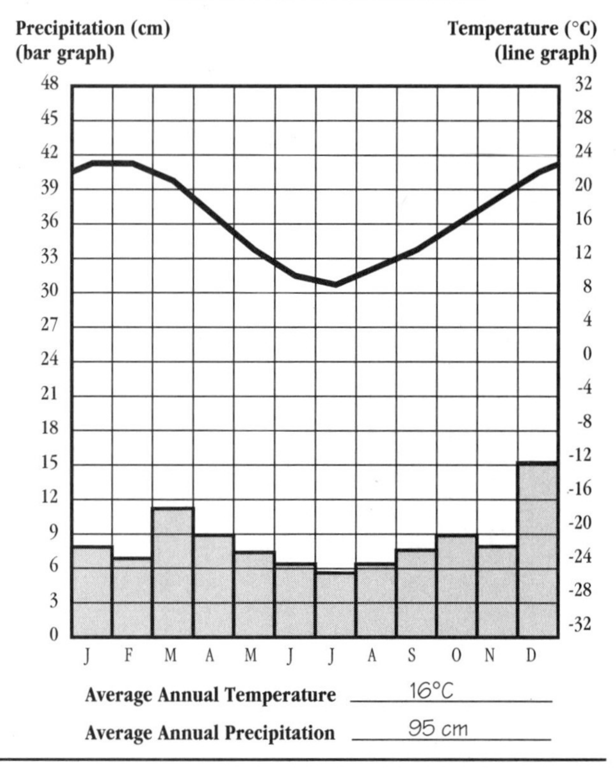

Average Annual Temperature 16°C

Average Annual Precipitation 95 cm

Climographs for Stations E through H

What Are the Characteristics of the River?

Focus:
Place Characteristics

In this lesson, students are introduced to the physical and human characteristics of a river environment. They learn to identify place characteristics in a river landscape and note how they change over time. Students learn to read topographic maps and identify features portrayed on such maps.

Perspectives and Geography Standards

Perspectives: Spatial and Ecological

Geography Standards Receiving Primary Emphasis

Throughout this lesson, students will learn:

1. How to use maps and other geographic representations, tools, and technologies to acquire, process, and report information from a spatial perspective
3. How to analyze the spatial organization of people, places, and environments on Earth's surface
4. The physical and human characteristics of places

Geography Standards Receiving Secondary Emphasis

During specific activities in this lesson—Teacher Presentation 3.1 and Student Information 3.1—students will learn about spatial change of a community over time (Standard 17); the effects of the physical environment on humans (Standard 15); change in an ecosystem (Standard 8); and human modifications of the physical environment (Standard 14).

Learner Outcomes

Students will:

1. Distinguish between physical and human place characteristics, particularly of a river.
2. Understand a selection of river-related vocabulary.
3. Recognize how place characteristics change over time.

4. Identify the information given on a topographic map and use a topographic map to understand the physical and human place characteristics of the project study area.
5. Discover relationships of phenomena by observing spatial patterns on a map.

Time

Five class periods of 40–50 minutes per period

BEFORE STARTING:	Assign Student Information 3.1: Place Characteristics of Rivers as homework
DAY 1:	Student Information 3.1: Place Characteristics of Rivers Student Activity 3.2: Place Characteristics—Vocabulary Teacher Presentation 3.1: Place Characteristics Over Time Assign Student Information 3.3: Topographic Maps
DAY 2:	Student Information 3.3: Topographic Maps Student Activity 3.4: Contour Lines
DAY 3:	Student Activity 3.5: Topographic Maps and Place Characteristics (with corresponding Teacher Guide)
DAY 4:	Student Activity 3.5: Topographic Maps and Place Characteristics
DAY 5:	Discuss Student Activity 3.5: Topographic Maps and Place Characteristics Student Assessment 3.6: Sketching a Map

Advance Preparation

Distribute Student Information 3.1 as homework before beginning this lesson. Prepare to supply students with Student Information 3.3; Student Activity 3.2 3.4, and 3.5; and Student Assessment 3.6. Make extra copies of Student Activity 3.4 so students can practice contour lines as needed.

Gather all necessary materials for this lesson. Prepare overhead transparencies of Teacher Figures 3-1 through 3-6, at end of Teacher Presentation 3.1 (Natural Landscape, Early Settlement, Growth of the Settlement, Flood, Rebuilt City, and Modern City) and of Island Map with Elevation Points (Figure 3-6 in Student Activity 3.4). Laminate topographic maps of the study area (one per two students, plus one for classroom use). (Alternatively, use photocopies of the topographic map portion covering the study area, one per student.)

Display in the classroom a copy of the topographic map of the study area, as well as other topographic maps that include terms and symbols not included on the study-area map. (Look for maps with a landscape very different from that of the study site. USGS staff may be able to suggest maps with the desired features. For contact information, see Appendix C.)

Materials

Teacher Presentation 3.1: Place Characteristics Over Time

For the teacher

transparencies of The Natural Landscape, Early Settlement, Growth of the Settlement, Flood, Rebuilt City, and Modern City (Teacher Figures 3-1 through 3-6)

overhead projector

washable marker

Student Activity 3.4: Contour Lines

For the teacher

transparency of Island Map With Point Elevations (Figure 3-6, in Student Activity 3.4)

Student Activity 3.5: Topographic Maps and Place Characteristics

Per class

topographic map of study area, displayed in classroom

additional topographic maps to show features not present on study-area map

Per two students

laminated topographic map of study area (7.5-minute series)

USGS pamphlet, *Topographic Map Symbols*

washable markers

2 rulers

Student Assessment 3.6: Sketching a Map

Per student

graph paper

ruler

Optional

colored pencils

Vocabulary

alluvial fan
area symbol
artificial levee
bank
bar
bench mark
bluff
boat ramp
braided river
bridge
contour interval
contour line or contour
cultural feature
cut bank
cutoff
landscape
line symbol
local relief
lock
map symbol
marsh
meandering river
natural levee
oxbow lake
physical place characteristics
point bar
point symbol
pond
quadrangle
rapids

dam	relief
dock	riprap
elevation	sediment
erosion	settlement
ferry	stream
flood	swamp
generalization	topographic map
gradient	topography
human activity	valley
human place characteristics	vegetation cover
index contour	velocity
interrelationship	waterfall
island	wetlands
lake	wing dam

Background for the Teacher

For information on obtaining topographic maps, the USGS pamphlet on topographic symbols, helpful classroom supplements entitled "What Do Maps Show?" and "Exploring Maps," and other USGS information, see Appendix C and pages xvi–xvii in the unit introduction. *The Language of Maps* by Philip J. Gersmehl (see Appendix C) is an excellent manual designed to introduce students to the language used to express spatial relationships.

Introducing the Lesson

1. Give students a brief overview of Lesson 3. Discuss Student Information 3.1: Place Characteristics of Rivers, and review student answers. (Sample answers are in Appendix B.)
2. Distribute Student Activity 3.2: Place Characteristics—Vocabulary. Instruct students to place a check beside each vocabulary term they can define. Ask for a show of hands to indicate their familiarity with the terms prior to reading Student Information 3.1. ("Were you familiar with 50 percent of the terms? 75 percent? 90 percent? How many can you explain to someone who knows nothing about them?")
3. Review the vocabulary list in Student Activity 3.2, referring to Student Information 3.1 and the Glossary as necessary. Ask for volunteers to share definitions and refine them if necessary. Assign the written part of Student Activity 3.2 as homework.
4. Present to your class Teacher Presentation 3.1: Place Characteristics Over Time.
5. Point to topographic maps displayed in the classroom. Explain that geographers commonly use topographic maps to study physical and human place characteristics of a location or region. Tell students that they will be working with topographic maps in the days to come, so they

need to understand the meanings of the symbols and other information that appears on these maps.

6. Assign Student Information 3.3: Topographic Maps as homework for the next class.

Developing the Lesson

1. Review Student Information 3.3, including student answers to questions. (Sample answers are in Appendix B.) Clarify as necessary.
2. Referring to the topographic maps on display in your classroom, explain the following information about topographic maps:
   a. Point out the map's name and date in the upper and lower right margins of the maps. Some maps have been photorevised since the first date stated; changes are shown in purple. Also point out the abbreviated legend at the lower right showing road classification.
   b. Ask students to recall the geographic latitude-longitude grid, and review it if necessary. Identify the latitudes shown on the top and bottom margins of the map and the longitudes shown on the left and right margins. (A 7.5-minute series USGS topographic map represents 7.5 minutes of latitude and longitude.) The coordinates on a particular map tell the *area* the map represents (rather than a point). To aid location, point out on a larger wall map the area included on the topographic map.
   c. Demonstrate the map scale, shown at the bottom of the topographic map. Review the concept of representative fraction. Use a ruler to show how to use the graphic scale at the bottom of the map to determine distances from one feature to another. (For specific teaching activities on representative fractions, see *Rivers Mathematics*.)
   d. Distribute the USGS pamphlet *Topographic Map Symbols* and review the color classifications and other symbols. Note that all features shown on the map are symbolic representations.
   e. Explain how contour lines are used to illustrate surface elevations. (For additional specific teaching activities on contour lines, see *Rivers Earth Science*.)
3. Have students do Student Activity 3.4: Contour Lines. Review student efforts informally as they work. If students are having difficulty, use the transparency (or a model copy) of Figure 3-6 from this activity to demonstrate the correct contour line. Remind students that the goal is to practice drawing contour lines until they can do so accurately, rather than to produce a product for a grade. As needed, provided students with clean copies of the handout so they can start over. When students are finished, have them compare their lines with those you draw on the transparency.
4. Distribute to pairs of students Student Activity 3.5: Topographic Maps and Place Characteristics, a laminated topographic map of the project area, and a washable marker. On the map, locate the field-study site. Have students complete this activity in class. (Use corresponding Teacher Guide.)

5. When students have finished labeling examples of vocabulary with a marker on their topographic maps, review the vocabulary list and their examples. You may have students point out on the display map the examples they found. For vocabulary words that do not appear on the study-area map, show examples on the other topographic maps on display in your classroom.
6. The next day, provide class time for the completion of Student Activity 3.5.

Concluding the Lesson

1. Review Student Activity 3.5 as a class. (See corresponding Teacher Guide). Ask for volunteers to read their statements for the Critical Thinking Questions. Invite others to analyze them critically.

Assessing the Lesson

1. As desired, collect and evaluate student answers for Student Information 3.1 and Student Activity 3.2 (answers in Appendix B) and Student Activity 3.5 (answers in corresponding Teacher Guide).
2. Have students predict possible changes to the landscape if selected features were hypothetically modified. For example, ask students how the construction or elimination of a levee or dam might affect a specific location. What effects would increased population have on the landscape?
3. Assign Student Assessment 3.6: Sketching a Map. The student handout contains the performance criteria; here is a suggested scoring rubric:

Scoring Rubric

Score	Expectations
0	No response attempted.
1	Incomplete effort. Some physical and human features are included, but major features may be omitted or not labeled. The map is generally inadequate to illustrate relative locations or directions. Mapping elements may be absent or incorrect.
2	Generally satisfactory description of items requested. Major features are included and labeled. Relative locations and directions are generally correct, although minor errors may be observable. Mapping elements are generally satisfactory.
3	Excellent description of items requested. Map accurately portrays locations and directions. Physical and human features are detailed and labeled. Mapping elements are accurate.

Extending the Lesson

1. Conduct further discussion, asking students to consider topics such as the following:
   a. How have physical features influenced how people have adapted to and altered them? Use examples to explain.
   b. Should agriculture be permitted on sloping ground?
   c. Should residences be permitted in a floodplain?
   d. What influence does extensive paving or covering of the earth's surface by highways, parking lots, driveways, and roofing have on natural drainage?
   e. What are the advantages and disadvantages of channeling rivers, building dams and levees, and dredging?
   f. How do altering river channels, dredging, and building dams and levees change the character of the river, its ecosystems, and river-related human activities?

TEACHER

PRESENTATION 3.1

Place Characteristics Over Time

Purpose

After students have become familiar with place characteristics in Student Information 3.1 and Student Activity 3.2, to show students how place characteristics change over time.

Background

Emphasize that recognizing change and understanding the processes that facilitate change are important in order to understand the geography of a location or region.

To make the most of this presentation, research the time period when your local river or stream area was first settled by Native Americans, by Europeans, and by others, as well as the times when similar places were settled in other parts of the country.

Materials

For teacher
transparencies of The Natural Landscape, Early Settlement, Growth of the Settlement, Flood, Rebuilt City, Modern City (Teacher Figures 3-1 through 3-6) at end of this Teacher Presentation
overhead projector
washable marker

Procedure

1. Tell students that you will show them a succession of illustrations of a river landscape that has changed as physical and human forces have influenced it. They should identify the place characteristics in each illustration. With each successive transparency, they should identify new place characteristics and any changes that have occurred.
2. Show overhead transparency of The Natural Landscape. Instruct students to call out features they recognize. *(Responses should include river, channel, bank, bar, bluff, cut bank, erosion, floodplain, gradient, island, local relief, waterfall, sediment, meandering river, swamp, tributary, valley, wildlife, wetlands, vegetative cover.)* If you wish, you or one of your students can label the features on the transparency with a washable marker.
3. Encourage students to think about the natural landscape by asking questions such as the following:
   - Are the features shown physical or human? *(Physical)* Are any human features visible? *(No)*

- Where is the river current fastest? *(Outer part of meander curve)* What evidence do you see that the current is slower on the inside of the curve? *(Slower current drops sediment on point bar.)*
- How did the waterfall develop? *(Stream flows over resistant rock and falls to level of eroded-away rock.)* Does such resistant rock appear elsewhere in the landscape? *(Yes; bluff at right, distant center)* Is it likely that these features are all part of the same rock formation? *(Yes, because they are about same elevation.)*
- How would you describe the river's gradient? *(Low or slight)* Why? *(A meandering river is characteristic of a floodplain with a slight gradient; no waterfalls or rapids occur in the main channel.)*

4. Before showing the next transparency, point out any features students have missed.
5. Show overhead transparency of Early Settlement. Ask students to name the human features that are shown in this transparency and have them identify any changes they see from the first transparency. *(The human features present for the first time include a house, ferry boat, road, wagon, and field for crops. Trees have been cleared; wildlife is less plentiful.)*
6. Encourage students to think about changes brought about by early settlement by asking questions such as the following:
   - What purposes are served by cutting down trees? *(Clears land for building and planting crops. The wood from the cut trees can be used for constructing houses and fences and for fuel.)*
   - Why was this particular site chosen for a human settlement? *(A need developed to provide transportation across the river.)*
   - Can you guess the approximate historic time this settlement might have been built based on the settlement of your local area? Do you think it was settled sooner or later than similar places in other parts of the country? *(Estimates will vary with time of first settlement in the local area and students' knowledge of history.)*
7. Before showing the next transparency, point out any features students have missed.
8. Show Transparency 3: Growth of the Settlement. Ask students to identify new features and changes in the landscape. (Responses should include: *more houses; buildings clustered primarily on one central street; larger buildings, possibly indicating the development of a commercial district; improved ferryboat landing; fewer trees; construction of a church; homes built across the river.)*
9. Encourage students to think about changes brought about by growth of the settlement by asking questions such as the following:
   - Why do people tend to cluster together in settlements? *(To be close to goods and services, for mutual protection, to meet social needs)*
   - What kinds of goods and services would you expect to find in a settlement of this size? *(Basic goods and services: providing for basic household needs, such as food, clothing, tools; meeting basic social and*

cultural needs through establishment of a church, a school, and perhaps a local government center)

10. Before showing the next transparency, point out any features students have missed.
11. Show overhead transparency of Flood. Lead students to understand that flooding is a natural process. Destruction of human settlements by flood waters usually occurs because the inhabitants settled in a floodplain and did not adequately adapt to the physical environment. Ask questions such as the following:
    - Is the flood a result of natural processes or human modifications? *(Natural)*
    - How could destruction of the community have been prevented? *(By establishing the community in a location above the floodplain; by building levees and floodwalls)*
12. Show overhead transparency of Rebuilt City. Ask students to identify new features and changes in the landscape. (Responses should include: *Natural changes as the result of flooding include cutoff of a meander and the beginning of an oxbow lake. Human modifications include levee, bridge, several rebuilt commercial and residential buildings, permanent roads, gravel mining, the old church in ruins, drive-in theater, removal of natural vegetation.)* Ask students questions such as the following:
    - What did humans do to alter the landscape in order to avoid flooding in the future?
    - What are the possible results of those changes?
    - How has land use changed compared with before the flood?
13. Show overhead transparency of Modern City. Ask students to identify in what ways the city has been built more, and the natural landscape altered more, in the time since the initial rebuilding. (Responses should include: *enlarged highways,many additional commercial and residential buildings including a residential subdivision, more extensive levee, new bridge, modification of abandoned channel to serve as boat anchorage, trees planted along bank, drainage of the swamp, paved parking lots, restoration of the old church, further removal of natural vegetation.)*
14. Ask students questions, such as the following:
    - As human activity at a location increases, what happens to the physical environment? *(It is altered to accommodate human needs by removing trees and other natural vegetation; using levees and floodwalls to restrict the river to its channel; replacing natural vegetation with managed vegetation; altering land surfaces by paving, mining, draining wetlands, or building embankments. As a result of these changes, wildlife in the area is reduced or eliminated.)*
    - How is the natural drainage system altered? *(Unnatural surfaces, such as pavements and rooftops, reduce absorption of water into the ground*

surface and cause increased runoff, which must be controlled by artificial systems to remove rainwater. The river is no longer free to change course. The floodplain is no longer available for overflow during times of flood. Artificial systems must be installed to drain excess water from land that was previously wetland.)

15. Make concluding or unifying statement about changes in place characteristics over time, such as, "Understanding change is vital to understanding geography. Relationships between humans and their environments that you can see at one point may change if some other feature changes. Though some changes occur suddenly, as in the case of a flood, others may occur more slowly and be harder to detect." Encourage students to consider place changes over time as they consider their local river or stream.

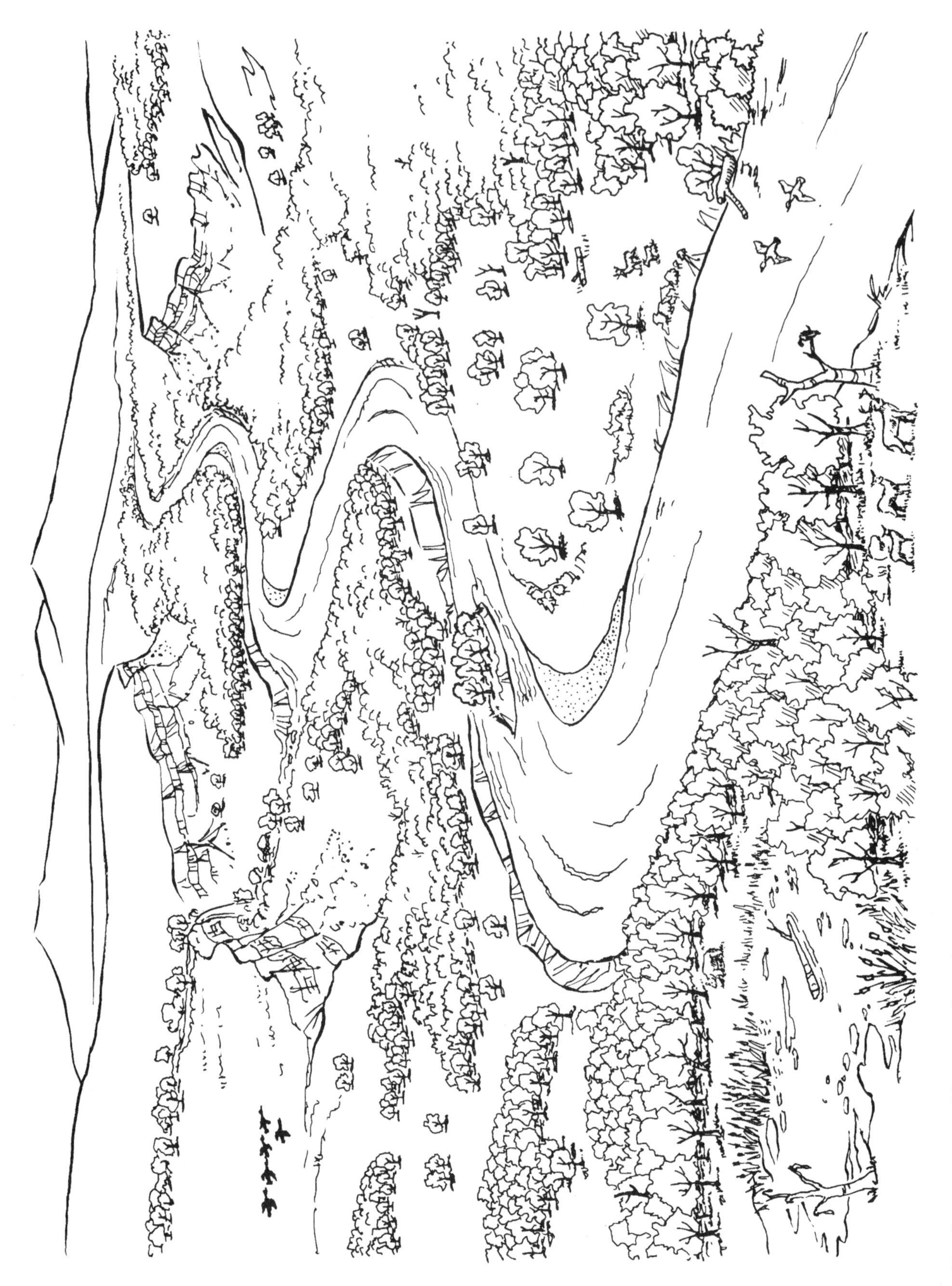

Teacher Figure 3-1: The Natural Landscape

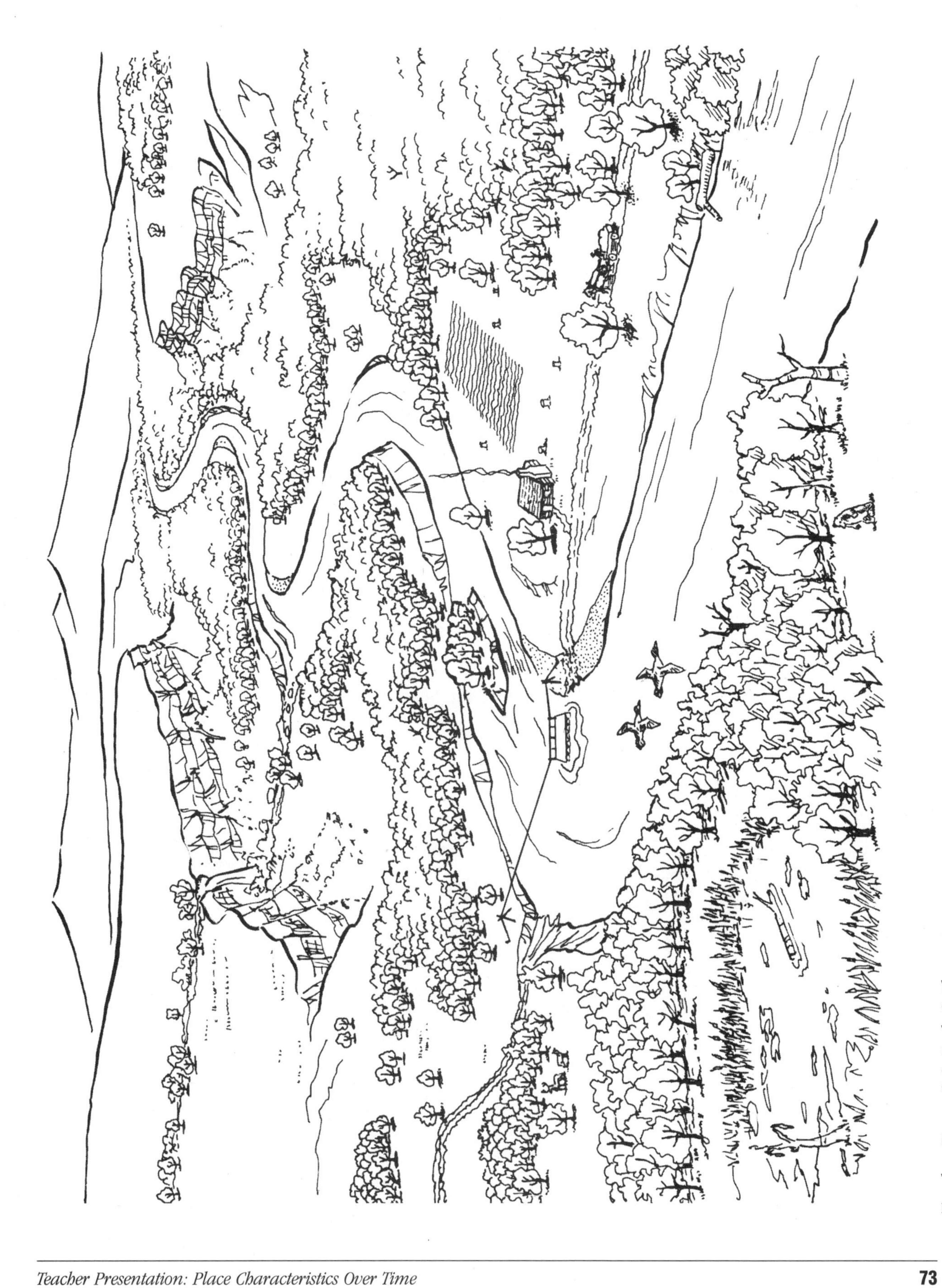

Teacher Figure 3-2: Early Settlement

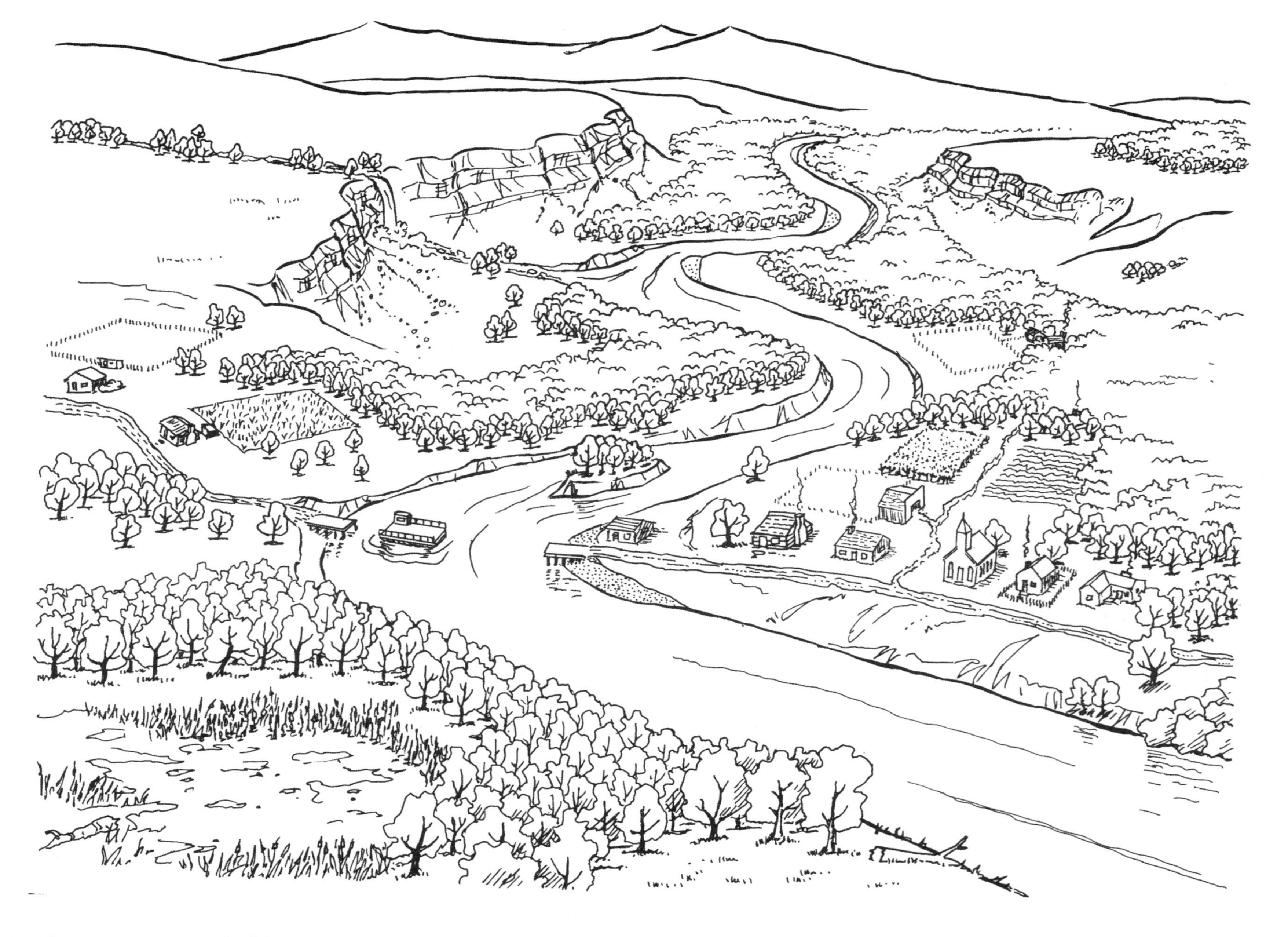

Teacher Figure 3-3: Growth of the Settlement

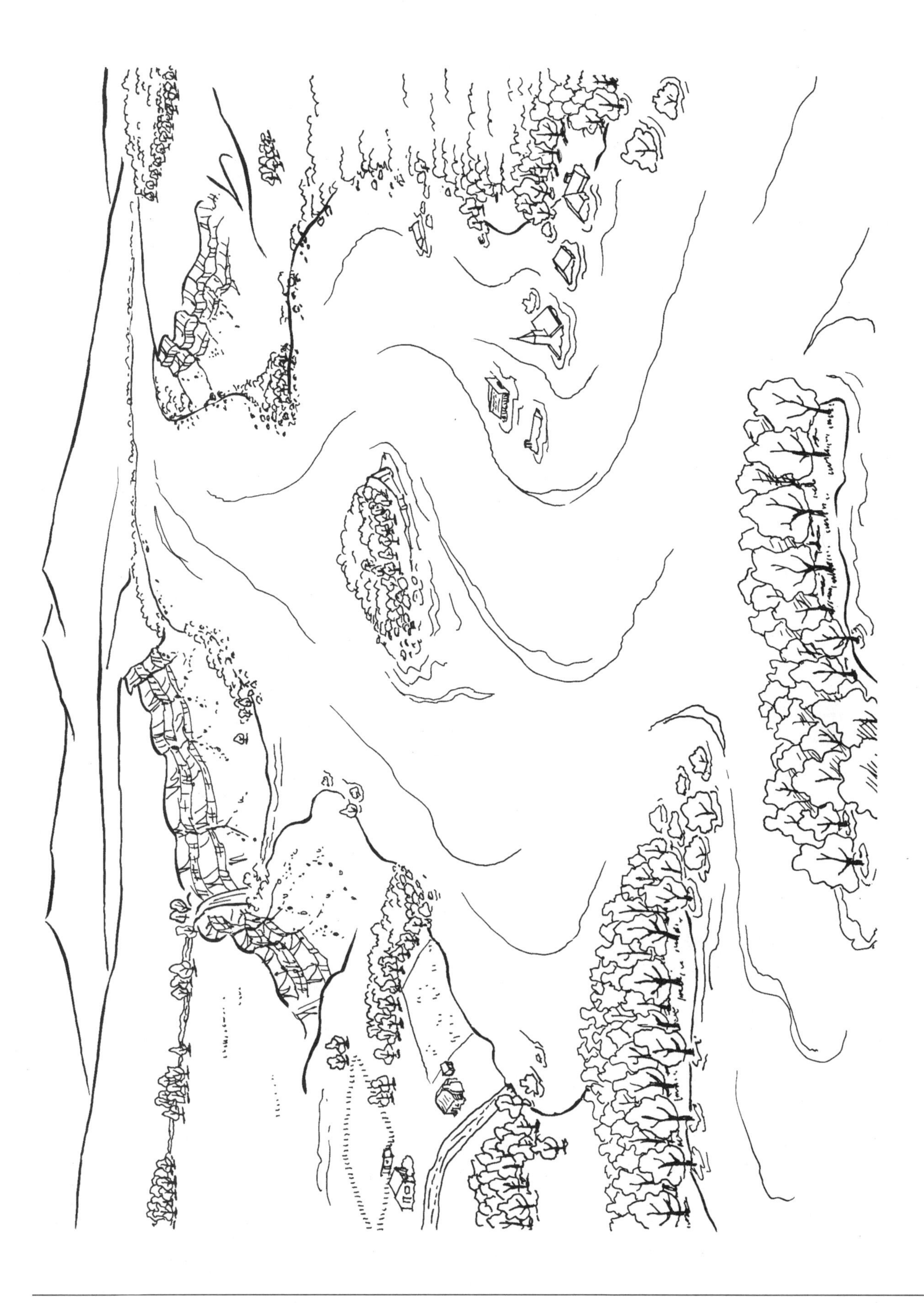

Teacher Figure 3-4: Flood

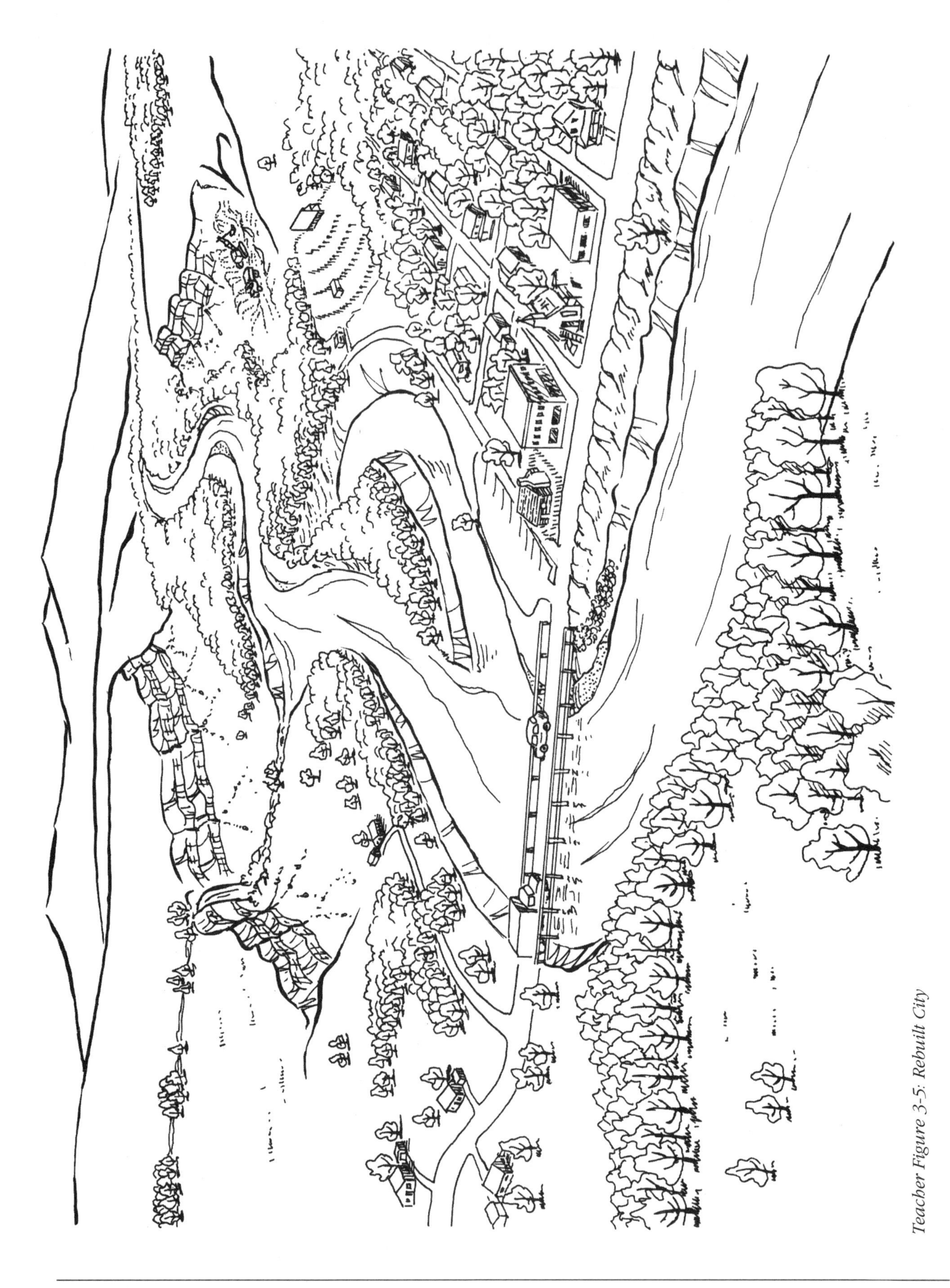

Teacher Figure 3-5: Rebuilt City

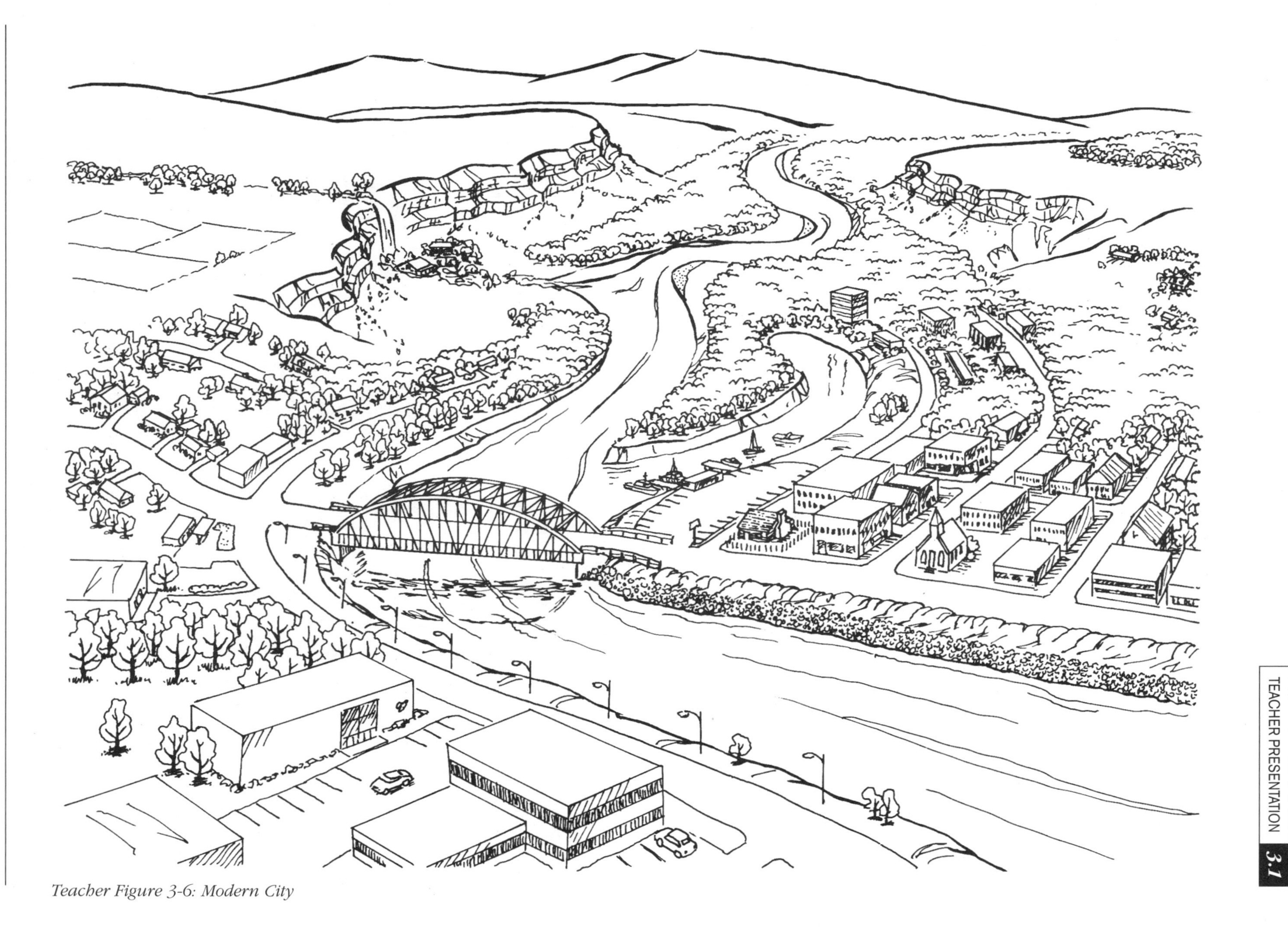

Teacher Figure 3-6: Modern City

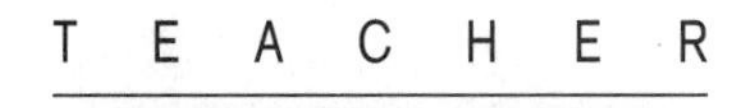

GUIDE

For Student Activity
3.5

Topographic Maps and Place Characteristics

Purpose

In parallel with Student Activity 3.5: Topographic Maps and Place Characteristics, to provide specific teaching steps that facilitate students identifying information provided by a topographic map and using a topographic map to make observations about physical and human characteristics of a landscape. (Suggested answers are in italics; notes for teacher are in bold.)

Background

Throughout this activity, students should use Student Information 3.1 and 3.3, and the booklet, *Topographic Map Symbols,* as needed.

Student Activity 3.5 provides a good opportunity for cooperative learning by having students work in pairs with the large topographic maps and symbol pamphlets. (If desired, this activity can be done on an individual basis, with each student having a map.)

This activity works quite well as an interactive, verbal exchange between teacher and students, with only the Critical Thinking Question and Keeping Your Journal segments used as written work. Alternatively, the earlier segments can also be used as written assignments. For materials list for this activity, see Student Activity 3.5 or Teacher Notes for Lesson 3.

Procedure

1. Read the pamphlet, *Topographic Map Symbols.* Review the color classification and other symbols illustrated in the pamphlet. Note that all the features shown on the map are symbolic representations.
2. On your topographic map, find as many examples as you can of the lesson vocabulary listed in Student Activity 3.2. Label each term on the laminated map with a washable marker. Not all are likely to be present on the map of your river study area. If you finish before others in the class, look for examples of remaining vocabulary on other topographic maps displayed in your classroom.

   When students have finished labeling their maps, review the vocabulary list and their examples with them. It may be instructive to tape a copy of the project map on the chalkboard and have students point out examples they found. For vocabulary words that do not appear on the project map, show examples on the other topographic maps displayed in your classroom.

3. See how much information you can obtain from your map by answering the Observation questions. Then complete the other segments of this assignment. For the segments your teacher asks for in written form, provide your answers on a separate piece of lined paper. Use complete sentences.

Observations

1. What is the name of the quadrangle shown on your map and on what feature is the name based?
The name of the quadrangle is given in upper and lower right margins of the map and will depend on the project area. The USGS commonly selects names of prominent features or features found near the quadrangle center.
2. Where is the area shown by your map? (Be sure to use N/S latitude and E/W longitude.)
_______ to _______ latitude, _______ to _______ longitude
The answer should denote an area marked by two latitudinal and two longitudinal lines, which are 7.5 minutes apart (if 7.5-minute series maps are used). Insist students use N/S latitude and E/W longitude.
3. When was your map published? *The date of publication is shown in upper and lower right margins. If the map has been photorevised, the date and changes on the map are shown in purple.*
4. What is the representative fraction of your map? *If the map is 7.5-minute series, the representative fraction is 1:24,000. If the map is a 15-minute series, the representative fraction is 1:62,500.*
5. What is the contour interval on your map? *The contour interval will vary with the amount of elevation difference in the quadrangle, but it is commonly 3 or 6 meters (10 or 20 feet).*
6. List the name of your project river. What tributaries of your river are shown on the map? *The names depend on the project river area. The project river and its tributaries may have nicknames or locally used names. Some small tributaries may not be named on the map but will have locally used names.*
7. List all other water or wetland features labeled on the map. *The other water or wetland features named on the map will vary. Some may be locally known but not named on the map.*
8. List all other natural features labeled on the map. *Answers will vary. Some features may be known locally but may not be named on the map.*
9. List all the human settlements (cities, towns) labeled on the map.
The list of settlements (cities, towns) labeled on the map will vary. Some may have come into existence since the map was published; some may have disappeared but still exist on the map; and others may have been recently incorporated or annexed by another community.
10. List all other human features labeled on the map. *The list of other human features labeled on the map will vary. Some features may have disappeared; others may have been added. Locally known names may not be on the map.*

Calculations

1. How many centimeters (inches) on the map represent one kilometer (mile) on the earth's surface? *1 km = 100,000 cm; therefore, 100,000/24,000 = 4.167 cm; 1 mile = 63,360 inches; therefore, 63,360/24,000 = 2.64 or approximately 2⅝ inches.*
2. How many kilometers (miles) on the earth's surface are represented by one centimeter (inch) on the map? *24,000/100,000 = 0.24 km; 24,000/63,360 = 0.379 mile.*
3. Measure the width of the floodplain.
   a. What is the widest measurement? ______ km (______ miles) Describe the relative location of the widest point.
   b. What is the narrowest measurement? ______ km (______ miles) Describe the relative location of the narrowest point.

   To measure the width of the floodplain, students may have used graphic scale or mathematical conversion. Allow for differences. Insist on the use of directions (N, NNE, NE and so on) and distances from some point labeled on the map.
4. If bluffs appear along the floodplain, calculate the local relief.
   a. Vertical distance from floodplain to top of bluff
   b. Describe the location of the points between which you measured. Use direction and distance from a point labeled on your map.

   Answers will vary depending on the points selected but should include cardinal and intermediate directions.

   Encourage students to identify the locations of their measurements. More than one measuring point may be usable.
5. What is the distance from school to the proposed river observation site?
   a. straight line ________km (________miles)
   b. by road ________km (________miles)

   To find the distance from school to the project observation site, students may have used the graphic scale or mathematical conversion. Allow for differences.
6. Compute how long it will take to drive from your school to your field site. Assume the driver obeys all traffic laws. *Answer will vary; calculate this as a class.*
7. Select a human feature and calculate its elevation. *The elevation will vary with the feature selected. Responses should cite the lower of the two contours between which the feature is located.*

Analyses and Conclusions

1. Why is 7.5 minutes of longitude less than 7.5 minutes of latitude? *7.5 minutes of longitude is less than 7.5 minutes of latitude because longitudinal lines converge at the poles. Latitude and longitude are equal at the equator, but longitude decreases with increased latitude.*
2. What changes have occurred since the map of your project area was published? Where and why have the changes occurred? Have the sizes of

human settlements grown or decreased since the map was published? How can you account for these changes? *Answers will vary. Settlements may have disappeared due to flooding, abandonment of a rail line, and so on. Updated maps show the addition of human activities in purple.*

3. What approximate scale is your map—small, intermediate, or large? How does its scale compare with scales on wall maps or city street maps? *A map with a scale of 1:24,000 is large scale; a map with a scale of 1:62,500 is intermediate. Both of the preceding scales are larger than the scale of a wall map and smaller than the scale of a city street map.*
4. What is the land surface like in those areas of few contour lines? Where are such areas found? How can one tell where the floodplain ends? Can you recall a time when the floodplain was covered with water? Is the floodplain used by humans? How? Are there buildings on the floodplain? Farms? Transportation lines? *In areas with few contour lines, the land is relatively flat, such as on a floodplain. The edges of a floodplain are indicated by lines that are closer together, indicating change in elevation. Human use of the floodplain will depend on how protected it is from flooding.*
5. How would you describe a wetland? Do you know of any water or wetland features not named on the map? *Wetlands are characterized by temporary or semipermanent water on the land surface, waterlogged soil, or poor drainage. Because of their temporary nature, some wetlands may not appear on the map.*
6. If ponds or lakes are present, are they natural or human-made? How can you tell the difference between a natural body of water and a human-made one? *Answers will vary. Human-made lakes are characterized by concrete or earthen dams or earthen levees. Some ponds and small lakes may not be distinguishable on the map as natural or human-made.*
7. Do you know the names of other features that are not named on the map? If so, what are they and why do you think they are not on the map? *Answers will vary, depending on how up-to-date the maps are. Some small features may be only locally known. Human features may have been added since the map was published.*
8. Using the information provided by labels, contour lines, and other map symbols, create a verbal description of the topography (physical and human features) shown on the quadrangle. *Answers should include descriptions of the general configuration of the land; water features; vegetation cover; human uses of the land, including concentrations of houses and other buildings, transportation lines, and other obvious modifications of physical features.*
9. Judging from the information given on your map, how have natural place characteristics influenced human activities? *Answers will vary but should include explanations of human attraction to or avoidance of physical features.*
10. What impact have humans had on the physical environment in your project river area? *Answers will vary but may include references to*

modifications such as damming, channelizing, or stabilizing rivers; removing vegetation cover (indicated by straight-line boundaries between forest and cleared land); and constructing individual buildings or cities and transportation lines.

Students may know many other examples from first-hand experience. Though such information may be of interest and worthy of development, emphasize that they should include here only information they can obtain from the map.

Critical Thinking Question

1. Look for relationships among any of the following features: relief, vegetation, wetlands, human settlements, transportation lines, locations of human features, streams, or others. Do two or more appear together? Does one appear where another is absent? Jot down some notes, citing specific examples and their locations for discussion. Select a pair of features, and write a statement describing how they relate. Give supporting examples, including mention of relative location. Your statement might assume the following form: "(A feature) tends to be found where (another feature) is located. Examples include the following: (Cite examples and locations.)" Repeat for a second pair of related features.

   Answers will vary with project area and student understanding. The following examples may apply:

   - *Rivers or streams tend to be located at elevations that are lower than surrounding locations.*
   - *Wetlands tend to be located in floodplains and along courses of streams.*
   - *Transportation lines tend to be located where there is minimal local relief.*
   - *Vegetation tends to be greater in areas of less human activity.*
   - *Floodplains tend to be of low local relief.*
   - *Wetlands tend to be in areas of low local relief.*

Have students offer reasons for the relationships. Insist that students include examples and relative locations of the examples. Ask for volunteers to read their statements and invite others to analyze them critically. Ask students if they can find examples that do not fit the statements. Ask if such examples make the statements invalid.

As appropriate, have students complete the Keeping Your Journal section.

Name

S T U D E N T
INFORMATION
3.1

Place Characteristics of Rivers

Once a geographer has identified and located an object of interest, she or he sets about to assess the characteristics that are similar to and different from those of other locations. Geography offers a unique perspective. Rooted in both the physical and social sciences, it serves as a bridge between the two. Through the study of geography, you can learn to recognize the influences and interactions of nature and humans on Earth's surface.

A **landscape** includes the total of human and physical features of a region. Elements of the landscape that result from natural processes are called **physical place characteristics.** This lesson will introduce you to the natural processes that shape rivers and provide you with a basis for recognizing the features of the landscape that are the result of these processes. You will learn to see rivers both as the results of natural processes and as the active agents of these processes.

Elements in the landscape that have been placed there by humans are called **human place characteristics.** Obvious visible features include houses and other buildings, transportation lines, agriculture, and mining operations. Human features of a particular location may assume a wide variety of forms, depending on the culture and level of technological development of the inhabitants. You can learn a great deal about the humans who live in a particular location by observing their structures and the ways they have altered the physical landscape to serve their needs and wants.

Geographers seek to understand the relationships among various physical and human features at particular locations and within particular regions. The arrangement of such features varies in kind and intensity, imparting a unique identity to each location and region.

DRIFTWOOD

"To a person uninstructed in natural history, his country or seaside stroll is a walk through a gallery filled with wonderful works of art, nine-tenths of which have their faces turned to the wall."

From *On the Educational Value of Natural History Sciences* by Thomas Henry Huxley

Physical Place Characteristics of Rivers

The natural process by which land surfaces are worn away by the actions of running water, wind, and ice is called **erosion** (ih ROH zhuhn). Because running water is erosive by nature, a river and its tributaries wear away the land over which they flow. The long, narrow depression created by the erosive force of a river or stream is called a **valley.**

A mature river is characterized by a flat floodplain immediately alongside the river channel. During times of **flood,** when the volume of water is greater than the channel can handle, excess water flows onto the floodplain.

When floodplains remain wet for long periods, bodies of water, such as ponds and lakes, may form. **Swamps** are wet spongy areas that are permanently waterlogged. **Wetlands** are temporarily saturated areas, the margin between dry land and open water. **Marshes** are poorly drained areas temporarily covered with water.

Rivers appear in all types of landform regions. In regions where the **relief**—or configuration of the land surface—is varied, rivers are fast flowing. In regions with gentle slopes, rivers flow slowly. When a river and its floodplain cut deeply into surrounding landforms, steep, prominent slopes, called **bluffs** or cliffs, are formed. Where bluffs are present, the **local relief**—or difference between adjacent high and low elevations—may be considerable.

Water flows if its source is higher in elevation than its mouth. The slope of a river or stream, the rate at which its elevation decreases, is called the **gradient** (GRAYD ee uhnt). Generally, gradient is greater in the headwaters at the source of a river or stream than at the mouth. As the gradient increases, the **velocity,** or speed of flow, of water increases.

The erosive action of water can alter the gradient of a river or stream. Where a river flows over rock that resists erosion and then over soft, easily eroded rock, **rapids** or **waterfalls** may form. The river wears away the softer rock, lowering the elevation of the downstream channel. The river water then flows over the resistant rock and falls to the lower elevation.

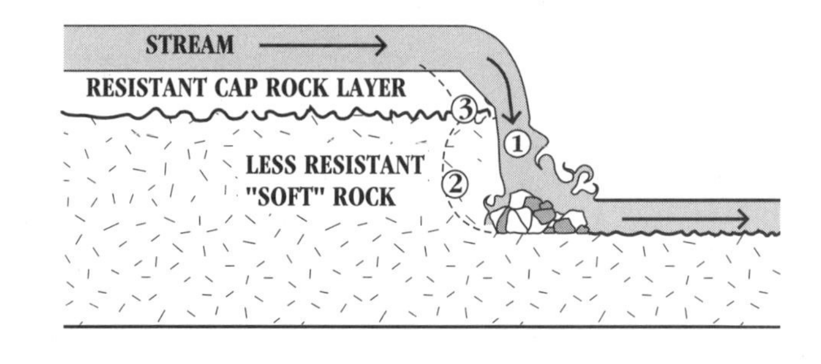

Figure 3-1: Formation of Waterfall: A stream undercuts the layer of resistant cap rock by eroding softer materials that lie underneath (1), causing an abrupt drop in the channel called a waterfall. As the water's action removes the more easily eroded underlying rock (2), support for the resistant cap rock is reduced. When the combined weight of the water and the rock become more than the structure of the cap rock can support, a portion of the cap rock breaks off (3) and falls to the base of the waterfall. Gradually, the waterfall recedes upstream.

The ability of a river to carry **sediment**—eroded materials transported by the river—is directly related to the velocity of the river. Rapidly flowing water can carry large amounts of sediment. When a river slows, it can no longer carry the same amount of sediment as it could when its velocity was greater. Where the water suddenly slows, the river deposits sediments as sand or gravel **bars.**

DRIFTWOOD

"As I sat there drinking water from cupped hands, I happen to look up and see on the opposite wall, a hundred feet above the floor of the canyon, the ruins of three tiny stone houses in a shallow cave. As is the case with many cliff dwellings, the erosion of eight centuries has removed whole blocks of rock which formerly must have supported ladders and handholds, making the ghost village now inaccessible."

From *Desert Solitaire* by Edward Abbey

A **braided river** consists of a network of channels flowing around many bars in the middle of the river. Braided rivers form when flow rates in a river fluctuate widely, as they do in rivers flowing through arid and semi-arid regions and in rivers fed by glacial meltwater. When a stream channel abruptly widens as it moves onto a plain or valley floor, the river slows and leaves fan-shaped accumulations of sand and gravel known as **alluvial fans.**

Where gradients are gradual, rivers tend to bend and curve. A **meandering** (mee AN duh ring) **river** is characterized by a curved channel that winds laterally across a floodplain. The river water moves most rapidly in the outer part of each curve. The outer part of the curve is called the **cut bank** because it experiences the most erosive action. As the cut bank erodes, the curve of the meander becomes more pronounced. The water slows on the inside of the curve where it deposits the eroded materials to form a **point bar.**

DRIFTWOOD

"Human subtlety...will never devise an invention more beautiful, more simple or more direct than does nature, because in her inventions nothing is lacking, and nothing is superfluous."

From *The Notebooks* by Leonardo da Vinci

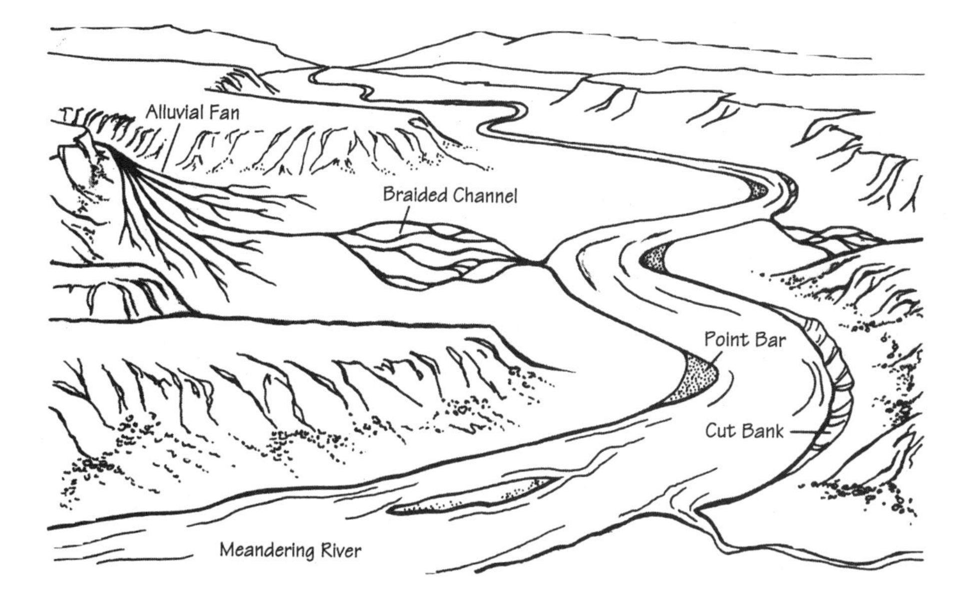

Figure 3-2: Features of a Meandering River

A river may cut through the narrow neck of land at the base of a sharply looped meander. The new, shorter channel is called a **cutoff.** An **oxbow lake** is formed when sediments fill the entrances to the former meander, isolating it from the new channel.

Natural levees are formed by sediments deposited along the banks of a river during floods. As the river overflows its channel, it drops heavier sediments along its banks, raising the elevation of the banks. Sediments also collect in the river bed and at the mouth of the river to form a delta. The natural cleansing action of floods scours the river bed of accumulated sediments.

Human Place Characteristics of Rivers

The natural environment poses challenges to human occupation. Depending on whether humans perceive the river to be a resource or a barrier, they make alterations in its natural character to serve their needs and wants. Such human interaction with the natural environment is one of the five themes of geography, which will be considered in greater detail in Lesson 5. Some information about the interaction between humans and their environment is presented here so that you can identify human features in the landscape.

Facilities constructed by humans while occupying a region or place are called **settlements.** Settlements along rivers may take the form of individual dwellings, towns, or cities. More populated settlements develop where bridges have been built to connect human activities on both sides of the river. Where the cost of a bridge cannot be justified because of infrequent traffic or great distance, ferry boats can provide a transportation link.

People sometimes construct **artificial levees** along the banks of a river to keep the river from overflowing its banks during times of flood. The purpose of an artificial levee is to protect settlements, transportation lines, and other economic activities on the floodplain. To provide additional dry land for human use, people also drain wetlands and install drainage channels and pumping stations to keep the land dry.

To accommodate river transportation, they may build dams and locks. **Dams** can provide deeper channels for navigation and create reservoirs of water for human use. Water falling from the dam can be used to generate hydroelectricity. **Locks** are chambers within a dam structure in which the water level can be adjusted so boats can pass from one side of the dam to the other.

A dam slows the flow of water in the river channel, increasing the amount of sediment that settles to the bottom of the river. If too much sediment collects in a channel used for transportation, maintaining an adequate navigation depth may require dredging. Navigation also sometimes may require straightening of the channel. To stabilize river banks, some projects add floodwalls or **riprap**—large stones or pieces of concrete placed on the sloping banks of a river to prevent erosion. To confine the navigation channel to a portion of the river, engineers may design artificial **wing dams,** low-lying structures extending perpendicularly from the bank in a shallow portion of the river. Docking facilities provide "tie-ups" for river vessels to load and discharge cargo. Smaller craft may enter and exit the river via **boat ramps.**

DRIFTWOOD

"These rapids, said Su-ling, standing at the window, had mainly been formed by a landslide in the second year of the reign of the Emperor Chia Ching of the Ming dynasty. Having been made so recently—only about three centuries ago—they were known, she said, as the New Rapids.

As we watched, three down-bound junks ran the rapids, and on the foredeck of each a river pilot stood like a black enormous bird on wide-planted feet, with arms raised and hands outspread, his huge sleeves and long gown flapping wildly in the turbulent air, but his head steady on his ramrod neck, transmitting signals to the frantic oarsmen by the lift of a finger on one hand, the folding of two fingers on the other hand, and so by sure economical movements of tiny sinews controlling the huge, awkward vessels in the fearfully entangling waters. No wonder they were such dominant, confident men as we had seen in the tearoom!"

From *A Single Pebble* by John Hersey

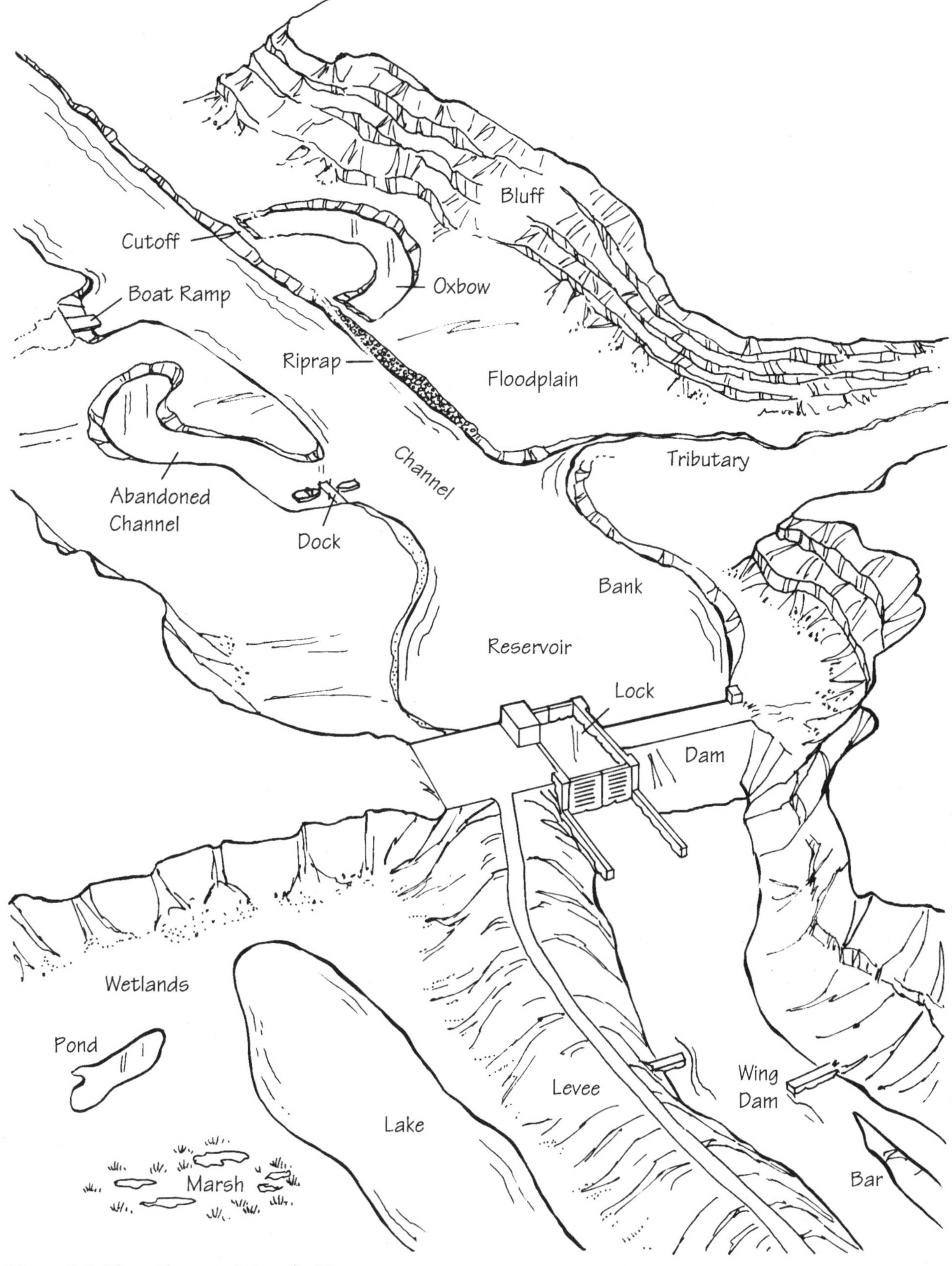

Figure 3-3: Place Characteristics of a River

Questions

Write your answers on a separate piece of paper. Use complete sentences (except in the lists for the first question).

1. Examine Figure 3-3 carefully. List each item shown in this figure, classifying it as either a physical place characteristic or a human place characteristic.
2. Explain why a river is the result of natural processes as well as an active agent of these processes.
3. Describe two physical place characteristics that are the result of the abrupt slowing of a sediment-filled river.
4. Give two examples of human modifications of rivers and explain what benefits they have for humans.

Driftwood

"Nature—that is, biological evolution—has not fitted man to any specific environment...His imagination, his reason, his emotional subtlety and toughness, make it possible for him not to accept the environment but to change it. And that series of inventions by which man from age to age has remade his environment is a different kind of evolution not biological, but cultural evolution."

From *The Ascent of Man* by Jacob Bronowski

Name

STUDENT ACTIVITY 3.2

Place Characteristics—Vocabulary

Purpose

To practice correctly using the terms used to describe physical and human place characteristics.

Background

Most of the terms you will use in this activity were described in Student Information 3.1 or in prior lessons. If you are unsure of the meaning of any of the terms, refer to Student Information 3.1 or look the terms up in your Glossary.

Vocabulary

alluvial fan
artificial levee
bank
bar
bluff
boat ramp
braided river
bridge
channel
cut bank
cutoff
dam
deposition
dock
erosion
ferry
floodplain
gradient
human activity
island
lake
lock
marsh
meandering river
natural levee
oxbow lake
point bar
pond
rapids
riprap
settlement
stream
swamp
tributary
valley
vegetation cover
waterfall
wetlands

Procedure

1. Using Student Information 3.1 and the *Rivers Geography* glossary as needed, mark with a "p" on the vocabulary list those terms that identify physical characteristics.
2. On unlined paper, draw a map of a hypothetical river environment. Plot the terms on the map where they would be likely to occur and label them. Do not include *erosion, deposition, human activity,* or *gradient* in your drawing. On the back of your drawing, write a brief explanation of each of these four terms as it applies to your drawing.

Topographic Maps

Topography (tuh PAHG ruh fee) refers to the human and physical surface features of an area. Maps that show the topography of selected areas of Earth, including the relief and physical and human features, are called **topographic** (tahp uh GRAF ihk) **maps.** Topographic maps have **contour lines** that show **elevation,** or the height of land above sea level or the depth below. A variety of symbols is used to show physical features—such as rivers, lakes, and forests—and human features—such as homes, roads, dams, railroads, and factories.

This information sheet will explain in detail what information you can expect to find on a topographic map and how to interpret this information. You probably will be using topographic maps that have been prepared by the U.S. Geological Survey (USGS). The area of Earth's surface represented by a USGS topographic map is called a **quadrangle.** Each quadrangle represents an area bounded by particular parallels of latitude and meridians of longitude.

Margin Information

The name of a USGS topographic map is stated at the upper and lower right margins of the map. The map name usually reflects the name of a prominent or centrally located feature on the map. The map series (7.5-minute or 15-minute) is shown at the upper right, and the date of information gathering and publication are provided at the bottom. At the center of each margin is the name in parentheses of the adjoining quadrangle.

Many older maps have been updated using aerial photographs. They contain the notation "photorevised" and the year the revision took place. Photorevised changes are shown in purple.

Scale

Large 1:24,000 scale maps are most useful for making a detailed study of an area. Such maps are part of the 7.5-minute series, in which each map shows an area of 7.5 minutes of latitude and 7.5 minutes of longitude. Intermediate scale maps in the 15-minute series have a representative fraction of 1:62,500 and show an area four times greater than the 7.5-minute series maps. The series is noted at the upper right, and the representative fraction is given at the center of the bottom margin. Also shown are graduated graphic scales for measuring miles, feet, and kilometers.

Symbols

As you will recall from Lesson 2, a map is a symbolic representation of the real world or a part of the real world. **Map symbols** of several types may be used to represent cultural or natural features. A **point symbol** designates a particular object at a specific location. For instance, a small black square may represent a house. A river mile marker (a short line with the notation "mile" plus a number to indicate distance from the river's mouth) is a point symbol that represents an idea. A **line symbol** may represent a physical object (such as a highway or power line) or an idea, such as latitude or longitude, a political boundary, or elevation (contour line). An **area symbol** may be used to show a physical object, such as a lake, or it may replace symbols for individual buildings with pink or gray shaded areas that indicate a densely populated urban area.

A standardized set of symbols and colors is used on USGS topographic maps. The legend for symbols appears in a pamphlet entitled *Topographic Map Symbols*. You should read that information carefully and keep the symbols handy when using topographic maps.

Location

A small insert map in the lower margin shows the quadrangle location relative to a larger region of the country. Latitude is shown at the top (N) and bottom (S) of the map, and longitude is shown at the left (W) and right (E); maps are oriented such that the top of the map is to the north. For example, the bottom of the Mosheim, Tennessee, 7.5-minute quadrangle is shown as 36 07'30", or 36 degrees, 7 minutes, 30 seconds (N latitude). The top margin is 36 15' (N latitude). Therefore, the map covers 7½ minutes of latitude. At the right margin is the designation 83 00' and at the left 83 07'30", or 7½ minutes of longitude. Notations in the margins divide the area every 2½ minutes. Beneath the name of the map in the lower right corner appears N3607.5-W8300/7.5, indicating the coordinates of the quadrangle's SE corner and the map series.

Elevation

A distinguishing feature of topographic maps is the graphic description they provide of the configuration of land surfaces, shown by a series of brown contour lines. Each contour line represents the level of the land above sea level, separating higher elevations from lower ones. For example, all points on a 440' contour line are 440 feet above sea level. Although contour lines in an individual quadrangle stop at the map's margins, they form continuous loops when placed next to maps of adjacent quadrangles. Within a closed loop are higher elevations, and outside are lower elevations, as shown in Figure 3-4. A special contour symbol is used to show depressions (as shown in the same figure).

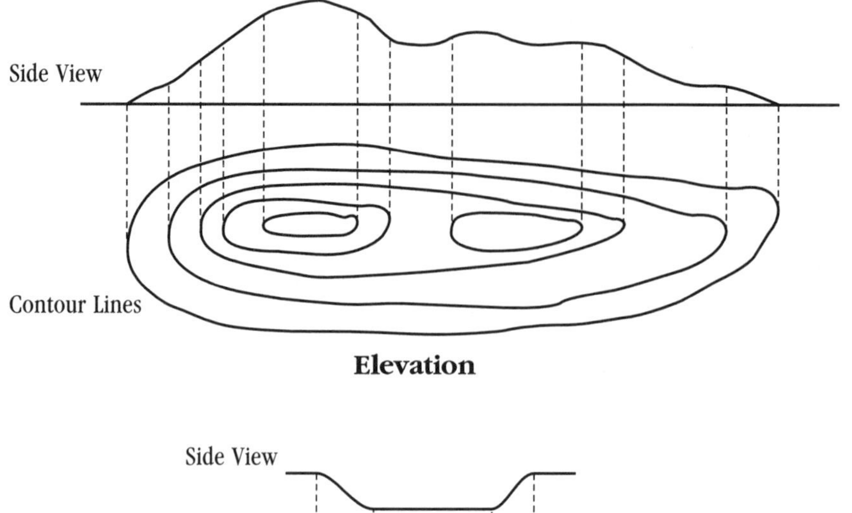

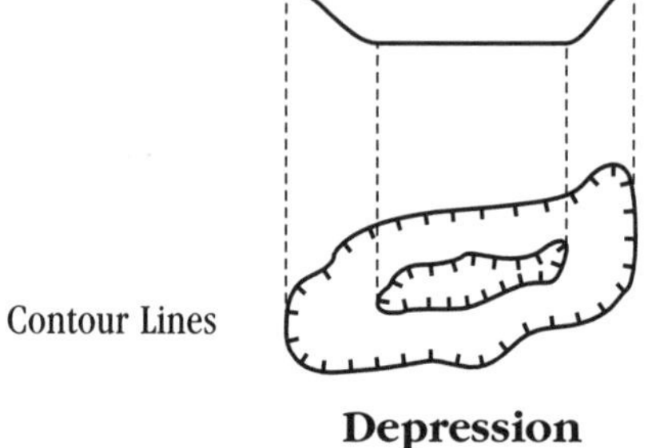

Figure 3-4: Contour Lines Show Elevation and Depression

Contours are useful for determining elevations of particular locations and for "reading" the land surface. Heavier brown lines, called **index contours,** appear every fifth line and have labeled elevations. Specific elevations can be checked at spot elevations (marked with an "X" and the elevation), water elevations (the abbreviation "ELEV." plus a number appearing on the water surface), or at vertical control stations as **bench marks** (shown as "BM X" plus a number). Elevations of bridges and crossroads are also commonly shown.

Local relief refers to the difference between the highest and lowest elevations of a particular area, whether it be a mountain range or a school campus. With practice, you can quickly get an idea of an area's topography by observing the **contour intervals,** the spacing or vertical distance between adjacent contour lines. Contours are close together on steep slopes and farther apart on gentle slopes.

When investigating the elevation of a river, remember that contours pass through points of equal value. So, because a stream bed has a lower elevation than its banks, contour lines run upstream along the banks in order to cross the river bed at the same elevation.

You can use contour lines to determine a river's gradient in two ways. First, to determine a river's gradient as the difference in surface elevation (in feet per mile), determine the different in elevation between two lines that cross the river; divide that difference by the stream distance between the two lines. For example, if one contour line is 1400' elevation and another line two miles away by stream distance has an elevation of 1320', then the gradient is 40 ft per mile (80 ft/2 miles).

Alternatively, to determine the gradient as a percent grade, use the following formula:

$$\text{Percent Grade} = \frac{\text{difference in elevation between points (in feet or meters)} \times 100}{\text{stream distance between points (in feet or meters)}}$$

$$= \frac{80 \text{ ft} \times 100}{10{,}560 \text{ ft}} = 0.0075 \times 100 = 0.75\%$$

Questions

Write your answers on a separate piece of paper. Use complete sentences.

1. What features do topographic maps show?
2. If you wished to make a detailed analysis of an area, which USGS series map should you use? How much area does each quadrangle in this series represent?
3. How are topographic map symbols used to represent both objects and ideas?
4. What are contour lines? How can contour lines be used to determine the gradient of a river?

Name

STUDENT ACTIVITY 3.4

Contour Lines

Purpose

To draw and interpret the contour lines on a map.

Background

The shoreline of a hypothetical island in a lake has been drawn for you in the figure on the next sheet, as has a small stream. The lake surface and its shoreline are 382 feet above sea level.

Using spot elevations ("X"), you are to fill in the contour lines to see what features are indicated by the map. As you do this, remember the following:

- Each contour line should pass through points of equal value.
- Contour lines are continuous, closed loops.
- Contour lines should never cross.
- Contour lines follow streams *upstream.*
- Contour lines are close together on steep slopes and farther apart on gentle slopes.
- Contour lines should separate higher elevation from lower ones.

Procedure

1. Using the point elevations plotted with "X" on the accompanying island map (Figure 3-6), use a pencil to draw contour lines on the figure at the following levels: 400 feet, 420 feet, 440 feet, 460 feet, 480 feet, and 500 feet. Complete the line at 400 feet before doing the 420-feet line, and so forth.

   You will notice that most of the spot elevations shown are intermediate between the elevations at which you will draw contour lines. You will need to approximate where the lines should be drawn. For an example, look at the figure below.

2. Place an "A" at the steepest slope, a "B" at the river's source, and a "C"at the river's mouth.

Analyses and Conclusions

Write your answers on a separate piece of paper. Use complete sentences.

1. Reading the map that you have drawn, describe in words the features it shows.
2. On your map, what is the local relief—the difference between adjacent highest point and lowest point—in metric measurement? (To convert to meters, multiply the number of feet by 0.30.)

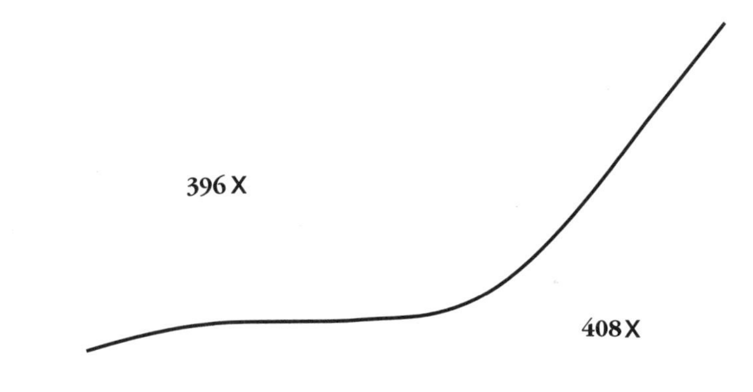

Figure 3-5: Drawing Contour Lines

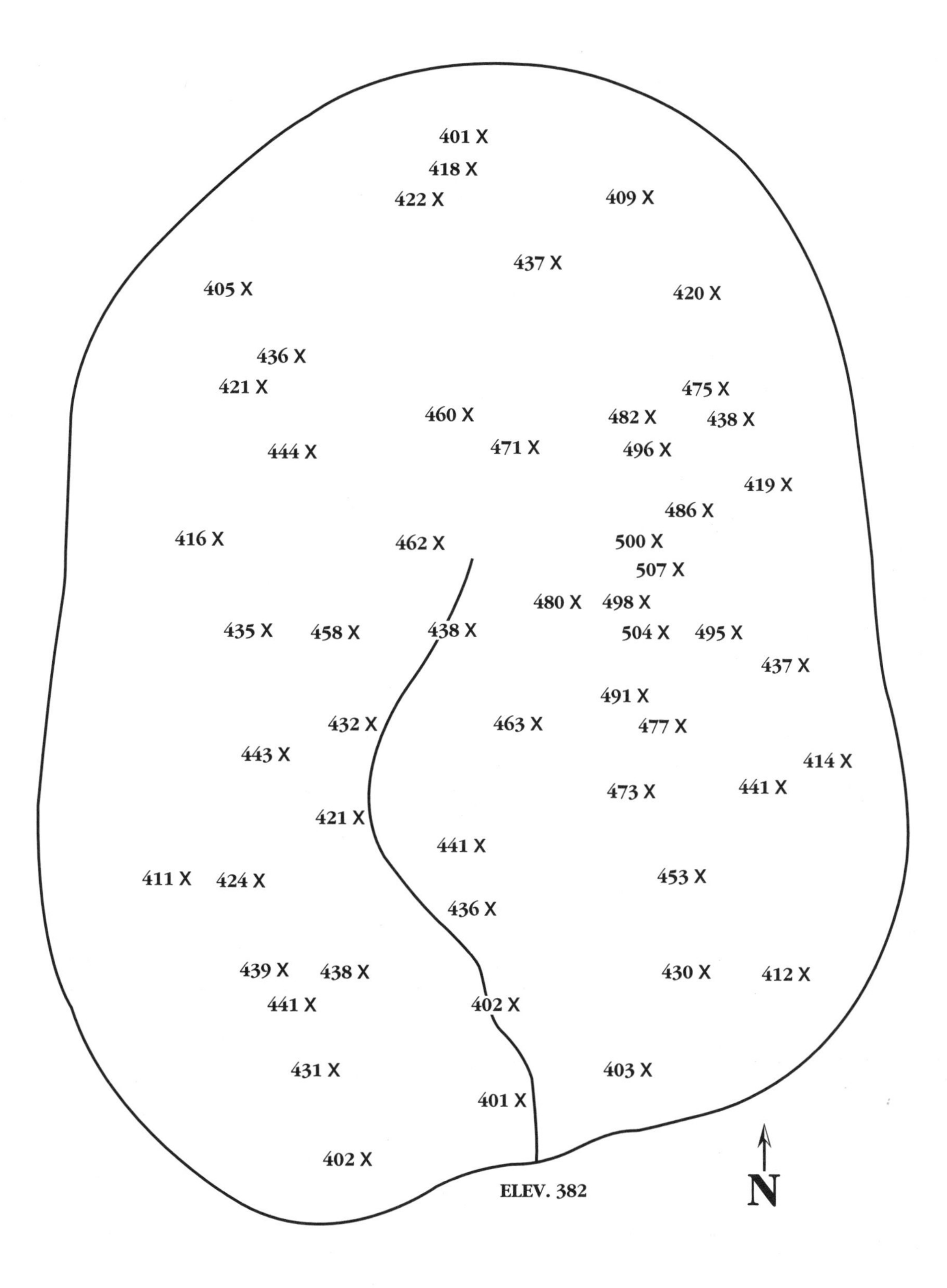

Figure 3-6: Island Map With Point Elevations

Name

Topographic Maps and Place Characteristics

Purpose

To identify information provided by a topographic map and to use a topographic map to make observations about physical and human characteristics of a landscape.

Background

Throughout this activity, use Student Information 3.1 and 3.3 and the booklet, *Topographic Map Symbols,* as needed.

Procedure

1. Read the pamphlet, *Topographic Map Symbols.* Review the color classification and other symbols illustrated in the pamphlet. Note that all the features shown on the map are symbolic representations.
2. On your topographic map, find as many examples as you can of the lesson vocabulary listed in Student Activity 3.2. Label each term on the laminated map with a washable marker. Not all are likely to be present on the map of your river study area. If you finish before others in the class, look for examples of remaining vocabulary on other topographic maps displayed in your classroom.
3. See how much information you can obtain from your map by answering the Observation questions. Then complete the other segments of this assignment. For the segments your teacher asks for in written form, provide your answers on a separate piece of lined paper. Use complete sentences.

Materials

Per class

- topographic map of study area, displayed in classroom
- additional topographic maps to show features not present on study-area map

Per two students

- laminated topographic map of study area (7.5-minute series)
- USGS pamphlet, *Topographic Map Symbols*
- washable markers
- 2 rulers

Observations

1. What is the name of the quadrangle shown on your map and on what feature is the name based?
2. Where is the area shown by your map? (Be sure to use N/S latitude and E/W longitude.)
   _______ to _______ latitude,
   _______ to _______ longitude
3. When was your map published?
4. What is the representative fraction of your map?
5. What is the contour interval on your map?
6. List the name of your project river or stream. What tributaries of your river or stream are shown on the map? Of what is it a tributary?
7. List all other water or wetland features labeled on the map.
8. List all other natural features labeled on the map.
9. List all the human settlements (cities, towns) labeled on the map.
10. List all other human features labeled on the map.

Calculations

1. How many centimeters (inches) on the map represent one kilometer (mile) on the earth's surface?
2. How many kilometers (miles) on the earth's surface are represented by one centimeter (inch) on the map?
3. Measure the width of the floodplain.
   a. What is the widest measurement? ______km (______miles) Describe the relative location of the widest point.
   b. What is the narrowest measurement? _____km (_____miles) Describe the relative location of the narrowest point.
4. If bluffs appear along the floodplain, calculate the local relief.
   a. Vertical distance from floodplain to top of bluff
   b. Describe the location of the points between which you measured. Use direction and distance from a point labeled on your map.
5. What is the distance from school to the proposed river observation site?
   a. straight line ______km (______miles)
   b. by road ______km (______miles)
6. Compute how long it will take to drive from your school to your field site. Assume the driver obeys all traffic laws.
7. Select a human feature and calculate its elevation.

Analyses and Conclusions

1. Why is 7.5 minutes of longitude *less* than 7.5 minutes of latitude?
2. What changes have occurred since the map of your project area was published? Where and why have the changes occurred? Have the sizes of human settlements grown or decreased since the map was published? How can you account for these changes?
3. What approximate scale is your map—small, intermediate, or large? How does its scale compare with scales on wall maps or city street maps?
4. What is the land surface like in areas having few contour lines? Where are such areas found? How can you tell where the floodplain ends? Can you recall a time when the floodplain was covered with water? Is the floodplain used by humans? How? Are there buildings on the floodplain? Farms? Transportation lines?
5. How would you describe a wetland? Do you know of any water or wetland features not named on the map?
6. If ponds or lakes are present, are they natural or human-made? How can you tell the difference between a natural body of water and a human-made one?
7. Do you know the names of other features that are not named on the map? If so, what are they and why do you think they are not on the map?

8. Using the information provided by labels, contour lines, and other map symbols, create a verbal description of the topography (physical and human features) shown on the quadrangle.
9. Judging from the information given on your map, how have natural place characteristics influenced human activities?
10. Using only information obtained from your map, what impact have humans had on the physical environment in your project river or stream area?

Critical Thinking Question

1. Look for **interrelationships**—two or more phenomena that influence one another—among any of the following features: relief, vegetation, wetlands, human settlements, transportation lines, locations of human features, streams, or others. Do two or more appear together? Does one appear where another is absent? Jot down some notes, citing specific examples and their locations for discussion. Select a pair of features, and write a statement describing how they relate. Give supporting examples, including mention of relative location. Your statement will be a **generalization,** a statement that expresses characteristics that exist in most, but not necessarily all, cases. It might assume the following form: "(A feature) tends to be found where (another feature) is located. Examples include the following: (Cite examples and locations.)" Repeat for a second pair of related features.

Keeping Your Journal

1. Using the map, select a specific vantage point from which you could view the river or stream and its environs. Still using the map, describe what you would see.
2. How can topographic maps be used when planning outdoor recreational activities?
3. Did you make any new discoveries about the area where you live when you looked at your topographic map? Write about what you discovered.

Name

Sketching a Map

Introduction

In this assessment activity, you will sketch a map that shows human and physical features and indicates relative locations and directions.

Procedure

1. Sketch a map showing the route you take home from school.
2. On your map, include and label:
   - home and school
   - route taken
   - roads and streets crossed en route
   - rivers, streams, and other water features, such as lakes, ponds, swamps
   - water-related human features, such as bridges, dams, levees, and boat docks
   - prominent physical features
   - human landmarks, particularly those that appear at points where the route changes direction
3. Be sure to include the elements of good mapmaking, such as the following:
   - title of map at the top
   - orientation (an N and an arrow pointing north) at the bottom
   - date and your name at the bottom
   - legend at the bottom explaining the symbols used
   - bar scale at the bottom showing distance on the map

Analyses and Conclusions

Write your answers on a separate piece of paper. Use complete sentences.

1. What area does your map cover?
2. What is the distance between your school to your home?
3. What physical and human features do you pass between your home and your school?

Performance Criteria

The successful map:

- Identifies human and physical features.
- Shows relative locations and directions.
- Displays the elements of good mapmaking.
- Is complete and accurate.

Materials

Per student

- graph paper
- ruler

Optional

- colored pencils

How Is Your River or Stream Site Related to Other Places?

Focus: Movement

In this lesson, students will learn about migration patterns and cultural diffusion. They will investigate the spatial distribution and spatial interactions of their ancestors and their cultures. They also will evaluate the historic and contemporary roles of the river or stream in their community.

Place
Location
Region
Movement
Human-Environmental Interaction

Perspectives and Geography Standards

Perspectives: Historical and Spatial

Geography Standards Receiving Primary Emphasis

This lesson primarily promotes mastery of the following standards:

9. The characteristics, distribution, and migration of human populations on Earth's surface
12. The processes, patterns, and functions of human settlement
17. How to apply geography to interpret the past

Geography Standards Receiving Secondary Emphasis

Several other standards receive secondary emphasis during specific activities. Specifically, community and family research encourages students to evaluate why people move (Standard 3) and to better understand the cultural elements of the community (Standards 4, 6, and 10). Students construct a map to show the movement of several generations of people (Standard 1), and map interpretation requires the identification of regions (Standard 5).

Learner Outcomes

Students will:

1. Assess the spatial distribution of class family members and their ancestors.
2. Describe and map family genealogy, including completion of a family tree.
3. Identify the cultural characteristics of the people in the local community or region.

4. Analyze the physical characteristics of the local area and identify what changes have occurred since the area was settled.
5. Discuss the relationships between the physical features and human features of the landscape.

Time

Three class periods of 40–50 minutes per period

DAY 1: Review Student Information 4.1: Genealogy Research and Community History
Review Student Information 4.3: Movement and Cultural Diffusion
Review Student Activity 4.2: Tracing the History of a Local Family (use corresponding Teacher Guide), Part A

DAY 2: Student Activity 4.2: Tracing the History of a Local Family, Part B

DAY 3: Student Activity 4.2: Tracing the History of a Local Family, Part C

Advance Preparation

As part of Lesson 1, distribute Student Information 4.1: Genealogy Research and Community History and Student Activity 4.2: Tracing the History of a Local Family (including U.S. and world base maps), so students have adequate time to do their genealogy research and complete their family trees. (For source of base maps, if needed, see National Council for Geographic Education, in Appendix C.) Assign students to complete Part A of Student Activity 4.2 by the first day of Lesson 4.

Distribute Student Information 4.3: Movement and Cultural Diffusion as homework before beginning this lesson.

Gather all necessary materials for this lesson. As you collect city, regional, state, national, and world maps for display, make selections that correspond to the genealogical data of your students. For example, if a large proportion of your students are of Asian descent, include a wall map of Asia.

Calculate the number of locations to be plotted, and obtain sufficient colored map tacks or pins. A class of 30 students plotting four generations would need to plot 450 locations (30 black, 60 blue, 120 green, 240 red).

Select books and magazine articles that discuss initial settlement and other historical periods of your community, as well as historic maps and photographs showing distinctive community architecture and street scenes. Place them on library reserve, photocopy them, or otherwise make them available for student use. As appropriate, prepare a list of such resources.

Materials

Student Activity 4.2: Tracing the History of a Local Family

Per student (already distributed during Lesson 1)

base map of your country
base map of world
markers—black, blue, red, green

Per student

15 colored map tacks or straight pins and construction paper squares:
- 1 black
- 2 blue
- 4 green
- 8 red

atlas

Optional (per student)

list of references about community history
set of short articles, historic maps, or photographs about community history

Per class

folders with photocopies of articles, photographs, and other information about community history
city, regional, nation, and world maps mounted on bulletin boards or similar surface

Vocabulary

ancestor
architecture
community
cultural diffusion
cultural landscape
cultural pluralism
culture
culture area
culture island
diffusion
distance decay
emigration
ethnic group
family tree
field study
genealogy
geographic past
immigrant
immigration
material culture
melting pot
migration
mosaic
nonmaterial culture
pull factor
push factor
sequent occupance
spatial interaction
subculture

Background for the Teacher

Rivers and streams have traditionally influenced where people settled. A look at human migration in North America during past decades and centuries highlights this pattern of settlement. Examining a map, students can observe that most major cities in the United States and Canada are situated on rivers, lakes, or ocean harbors. In this lesson, students will observe what factors, including rivers and other bodies of water, have influenced human migration in their own families and communities. Rivers and lakes have also influenced the style of settlements and choices of occupation. Finally, students can learn how the cultures of immigrants have influenced the community and its inhabitants.

A good resource for this lesson is *Historical Atlas of the United States,* edited by Wilbur E. Garrett, published by the National Geographic Society (see Appendix C.)

Introducing the Lesson

1. Student Information 4.1: Genealogy Research and Community History and Student Activity 4.2: Tracing the History of a Local Family should already have been distributed to students during Lesson 1.
2. Give students a brief overview of Lesson 4. Before starting Lesson 4 classroom activities, assign Student Information 4.3 as homework and remind students to also complete Part A of Student Activity 4.2.
3. Discuss Student Information 4.1. (Answers are in Appendix B.)
4. Have students discuss and answer questions for Student Information 4.3: Movement and Cultural Diffusion. (Answers are in Appendix B.)
5. Review Part A of Student Activity 4.2 (student family trees) to make sure students are ready to proceed with the rest of Student Activity 4.2. (Use corresponding Teacher Guide.)

Developing the Lesson

1. Discuss information resources available about community history. Allow time for library or classroom research so students can complete Part B of Student Activity 4.2.
2. Review the Observations and Analyses and Conclusions questions in Student Activity 4.2. (Typical responses are in the corresponding Teacher Guide.)
3. Have students complete Part C of Student Activity 4.2, including displaying their genealogical data on large city, regional, U.S., and world maps, then answering the Critical Thinking Questions.

Concluding the Lesson

1. Conduct a classroom discussion based on students' answers to the Critical Thinking Questions in Student Activity 4.2.

Assessing the Lesson

1. Direct students to write a creative journal entry. Ask them to assume the perspective of the area's first settlers and describe why the location was a desirable one for settlement, making specific reference to the local river or stream. The journal writing may take the form of a letter, a diary entry, or a narrative.

Extending the Lesson

1. Gather census data about the ethnic makeup of your community. Using these data, have students graph or map the locations of various ethnic groups in your community.
2. Display a large-scale map of your local community in your classroom, or distribute individual copies to students. Have students identify various land uses (residential, commercial, industrial, or public) and assess the locational relationships and advantages or disadvantages of each. The following questions may be used to promote class discussions:
   - Why do certain commercial activities tend to be found in central business districts while others are located at the edges of the community? How do transportation resources influence the location of commercial activities?
   - Are older and newer businesses located in different places? Why?
   - Is the community growing or declining? Why? If the community is growing, where is the growth taking place and why is it occurring at that location?
   - What factors outside the community have influenced the growth or decline of the community?
   - Are efforts needed or underway to revitalize sections of the community?
   - How has the river or stream influenced land use and transportation?
3. Have students prepare a graph to illustrate the historic growth of the community population. Then ask the following questions:
   - What has been the pattern of growth?
   - Did the population grow faster at certain times than at others? How have local or national events affected the rate of growth or decline?
   - What is the recent trend? Is the trend likely to continue? Why or why not? What internal or external developments are likely to change the pattern?

TEACHER

GUIDE

For Student Activity
4.2

Tracing the History of a Local Family

Purpose

In parallel with Student Activity 4.2: Tracing the History of a Local Family, to provide specific teaching steps, sample answers, and suggested discussion topics that facilitate students learning how family and community histories relate to patterns of human migration and cultural diffusion. (Suggested answers are in italics; notes for teacher are in bold.)

Background

Have students start this research activity as soon possible, preferably during Lesson 1, to allow adequate time to collect data. Be sensitive to students who may be adopted or living in foster homes. If any students are not comfortable with researching their own families, make sure they have other options.

Encourage students to make research as far back as resources and time allow. Grading results for completeness and effort rather than for content encourages students to view their efforts in terms of their connection to studying geography. If desired, use student answers to Critical Thinking Questions as assessment tools.

Procedure

PART A. Making a Family Tree

1. Obtain the names of three generations of family members and their birthplaces (country, state, city), going back to great grandparents. Enter this information on the sheet provided with this activity to create a family tree. If you cannot find out all this information, include notes indicating the steps you took.
2. When you have completed your family tree, plot the birthplaces of the family members on maps of the United States and the world. Use a black marker to plot the place where you (or the person whose family you researched) were born. Use a blue marker to plot the birthplaces of parents, a green marker for grandparents, and a red marker for great grandparents.
3. Answer the questions in Observations under "About the Family."

PART B. Learning About Local History

3. Conduct research to find out about your community, using the resources available in the classroom, school library, public library, and among individuals, as guided by your teacher. Use the information you gather to answer the rest of the questions in the Observations and Analyses and Conclusions sections of this activity. Write on a separate piece of lined paper; use complete sentences.

Distribute a list of references about community history, if you have prepared one. If possible, make available to the students in the classroom or school library folders with other community history resources. Discuss other sources of such information. Allow library time for students to do this research.

PART C. Displaying Local Patterns of Immigration

4. When requested by your teacher, pin map tacks (or straight pins with squares of colored paper) on the designated country and world maps to represent the birth places of the individuals identified on the family tree you have made. Follow the colors on the legend as indicated by your teacher, such as:
   black pin—you or subject of study
   blue pins—parents
   green pins—grandparents
   red pins—great grandparents
5. Answer the Critical Thinking Questions. On your separate piece of lined paper, use complete sentences.

Observations

A. About the Family

1. When did the family you researched first move to your local community? *Answers will vary depending on your community and the composition of the student population.* **Exact dates, previous locations, and reasons for movement may not be known. Encourage students to try to match historic periods with local or regional activities or opportunities that might encourage immigration.**
2. Where did the family live before they moved to your community? *Answers will vary depending on your community and the composition of the student population.*
3. Why did they move to your community? *Answers will vary but may include improved employment opportunities, better schools, nicer housing, a more reasonable cost of living, or the presence of relatives or members of an ethnic population.* **Encourage students to think about why people move.**
4. What cultural practices in the homes of this family can be traced back to other parts of the country or world? Describe what you know of their origin. *Answers will vary.*

B. About the Community

Most communities have written histories that reveal dates of settlement and descriptions of early settlers. If such information is not available, encourage students to make generalizations based on the region in which the community is located. If the settlement of a nearby city is better documented, students may use that as an example if necessary.

5. When was your community first settled? *Answers will depend upon your community.*
6. What country or region were the first settlers from? Where were later settlers from? *Answers should reflect the different ethnic groups that migrated at various times. Native Americans should not be overlooked.*
7. Does your community sponsor celebrations, customs, traditions, or holidays (other than legal holidays) that emphasize the unique character (cultural religious, occupational, or recreational) of the community and the people who live there? Explain. *Most students will be able to name important celebrations. Many smaller events may be known to only a few of the students.* **If your community has few celebrations, encourage students to think of examples from other communities.**

Analyses and Conclusions

1. Why did people select your site for settlement? Did the location offer advantages for transportation or agriculture? *Answers will vary, but should be based on available historical accounts of the community's first settlers.* **Some influences on settlement, which you may wish to discuss further with your class, include the following:**
   - **land grants, the Homestead Act**
   - **railroad construction (establishment of water stops and stations, land grants to railroad companies and subsequent sale of lands to settlers)**
   - **general configuration of the land surface**
   - **transportation lines, crossroads of trade**
   - **urbanization and the growth of suburbs**
   - **planned communities and communal settlements**
   - **development of fencing (particularly barbed wire), windmills to tap groundwater sources, and the steel plow**
   - **discovery and development of mineral resources**
   - **establishment of administrative centers, such as county seats and state capitals**
   - **military installations**
   - **service centers providing basic goods and services to people in settled areas**
2. What role did the river, stream, or other waterway play in making the site attractive for settlement? Was the water used for transportation, as a water supply, or as a source of food? Did it help connect the site to other places, or was it a barrier? *For communities located on a waterway, multiple reasons are probably applicable. For communities that are some distance from the water, its role will assume less importance.* **You may ask students to consider the importance of the river in the settlement and development of cities such as Minneapolis–St. Paul (upper terminus of Mississippi River navigation), Pittsburgh (at the junction of the Monongahela and Allegheny Rivers, which form the Ohio River),**

or New York (transshipment point for land, river, and ocean transportation).

3. What role in the community does the river, stream, or other waterway play at present? Does it serve economic, recreational, or other needs? *Larger, navigable rivers are probably used to transport goods. Smaller rivers may be recreational centers for boating and fishing. Rivers also may be used to supply water for communities.*
4. List the human features and activities in your community that would not exist if it were not near the water. *Answers will vary but may include shipping terminals, boat sales and services, bait and tackle shops, riverfront shops and parks, bridges, levees, dams, and the jobs associated with them.*
5. Has the river, stream, or waterway been modified to serve human needs? Explain. *Answers should include any human-built features, such as bridges, levees, floodwalls, riprap, boat docks, boat ramps, locks, dams, and wing dams, channelization and dredging of the river for water navigation, boat anchorages that have altered the shoreline, and water treatment (if the water is used as a water supply.)* **Even in communities not located on a river, encourage students to find modifications that accommodate drainage, such as culverts, drainage ditches, and storm sewers.**
6. Has the physical form of the present-day community been influenced by the river, stream, or waterway? What adaptations have been made? *Answers will vary. For many communities, answers include: the central business district is near the river's edge, reflecting the earliest settlement area of the community; streets conform to the river's shape; bridges have been constructed to serve major streets; structures have been built to protect buildings from the river. Communities located some distance from the waterway may not have been directly influenced by it.*
7. Does architecture reveal changes that can be identified with particular periods of time? With particular cultural influences? List examples, including dates of construction and styles of architecture. *Different styles and cultural influences may be obvious in some communities. In other communities, distinctions may be difficult to make.* **Encourage students to compare the community's most recently constructed buildings with older ones. Ask them to compare architectural styles, building materials, locations, functions, and placement on lots.**

Critical Thinking Questions

Once students have finished using pins to mark their family-tree birthplaces on the class map, provide time for them to analyze the results and record their observations.

1. Briefly describe the patterns created by the plotted birthplaces of you and your classmates. *Patterns will depend on the mobility of people in the community. In many cases, students' birthplaces will tend to cluster around the community. In other cases, the pattern may be geographically widespread.* **Ask students to consider the factors that might**

contribute to the patterns they observe. Have them comment on the expected patterns of communities that are relatively new, are located near a military base, or are college or university towns.

2. What patterns are observable from the plotted birthplaces of the different generations: parents, grandparents, and great grandparents? Are these patterns similar to the pattern for your class? Why or why not? *Patterns tend to be more dispersed among older generations. Thus, the pattern displayed by the grandparents' birthplaces may be very different from the pattern of students' birthplaces.* **Remind students of some of the reasons for the differences: (1) The United States has been called a "Nation of Immigrants," and the farther back you go, the more likely you are to find an ancestor who is an immigrant. (2) The United States is characterized by a highly mobile society. As economic and social conditions change, people are likely to move to places that offer employment and other opportunities.**
3. If the birthplaces of more remote generations (earlier than great grandparents) were plotted on the maps, would you expect the pattern to be similar to or different from the patterns created by the birthplaces of more recent generations? If different, in what ways? Why? *More remote generations would differ markedly from present generations. (See also expected responses for Question 2.)* **Point out that such a map might reveal a clustering in another country or region. New immigrants tend to communicate with friends, families, or others who live in their previous homeland and who eventually may decide to immigrate.**
4. If the birthplaces of the next generation (your future children) were plotted, do you suppose the map would assume a similar or different pattern? Why? *Answers will be purely speculative. With increased mobility due to increased communication and transportation, present-day students may be widely dispersed by the time they begin to raise families. Nevertheless, a strong concentration of the next generation's birthplaces will probably occur in the community where students were born or spent most of their childhood.*
5. The concept of **distance decay** states that the frequency of occurrence of a phenomenon tends to decrease as the distance from a central point increases. Explain why people tend to live in or near the place where they were born. What causes people to move away from their birthplaces? *People tend to remain close to family, friends, and the environment with which they are familiar. They may move to new locations for a variety of reasons, especially the pull of better economic opportunities.* **Remind students of push and pull factors as influences on migration. Ask them to identify factors that might make them decide to move.**

6. How is your community unique? What activities or programs could be established to emphasize its unique character? *Students may suggest physical or human features of the community or nearby area worthy of celebrating.* **Encourage students to ignore the carnival rides, parades, and other amusements that accompany community celebrations, focusing instead on activities that encourage people to better appreciate the community's unique qualities.**

   As appropriate, have students complete the Keeping a Journal section.

Name

STUDENT
INFORMATION
4.1

Genealogy Research and Community History

Most people living in the United States either moved here from other countries or are the descendants of people who moved here from other countries. Each group of newcomers brought its own **culture,** or unique qualities, such as language and customs, that set those individuals apart from other groups. Each group has left its cultural imprint on the human landscape.

Communities

A relatively permanent cluster of persons who have gathered to satisfy basic needs and wants is known as a **community.** Many inhabitants of a community may share a similar cultural background. As a result, they tend to create a human landscape that is culturally familiar, or resembles the one from which they came. Thus, the **architecture** (AHR kuh tehk cher) (design and style of buildings), street patterns, celebrations, customs, and other elements in the human landscape often reflect the cultures of the groups who settled there. Different styles of architecture can indicate historical changes in the community. Many of the older buildings may have been replaced with more modern structures, which in turn may have been replaced with newer styles.

Using a combination of geography and history, a geographer can use cultural clues in the human landscape to reveal the changes that have occurred in a particular location or area during various periods of development. This **geographic past** is revealed by studying one place at a particular time—say, New Orleans in 1740. When additional time periods are studied to show how the landscape has changed over time, it is called **sequent occupance** (SEE kwuhnt AHK yuh puhns).

DRIFTWOOD

"The Mississippi River towns are comely, clean, well built, and pleasing to the eye, and cheering to the spirit. The Mississippi Valley is as reposeful as dreamland, nothing worldly about...nothing to hang a fret or worry upon."

Mark Twain

Genealogy

Genealogy (jee nee AHL uh jee) is an account of a person's descent from an ancestor or ancestors. An **ancestor** is one from whom an individual is descended and who is more remote than a grandparent. A genealogy often is recorded in the form of a **family tree,** which shows family members and their relationships to one another.

Many people research their family trees because they are curious about their ancestors. Often, they are able to connect their ancestors to historic events. In doing so, they find out about the history of the people who make up their community and their nation. They discover how people earned their livings, how people cooperated and associated with one another, and how people interacted with their environment.

Later in this lesson, you and your classmates will do family trees, then use the genealogical data you collect in order to create a map showing the locations of the birthplaces of people in your family trees. From this map, you should be able to identify patterns of movement of people who live in your community. These patterns will give you a better understanding of your community and its residents.

Researching a Family Tree

If you are going to study the genealogy of your own family, then the most important sources of information will be living family members and records of past family members. Some families have one or more "unofficial" historians who are able to document many generations of a family. Other families may have written histories.

Instead of your own family, you may research the genealogy of a friend, a neighbor, or some other member of your community. You probably will want to contact the person whose genealogy you will be researching. If the person's family can be traced back to the founders or early settlers of your community, you probably can find all the necessary information at your local library or historical society. For more extensive research, the Church of Jesus Christ of Latter-Day Saints (Mormons) maintains genealogical libraries in larger communities.

Researching Your Community

Later in this lesson, you will also research your community history. Your school library and your local public library are excellent starting points for such research. Your local library may have a special section reserved just for such historical materials. The main branches of public libraries also usually have a collection of newspapers available on microfiche.

You can also obtain information from citizens of your community, particularly those who have been residents for many years. Your teachers, parents, or neighbors may be able to help you identify individuals to contact and interview. Before conducting an interview, you should research the questions you want to ask and have these questions approved by your teacher.

Your personal observations of your community may reveal additional information about its historical development, its relationship to the river and other physical features, and the people who live there. Such firsthand gathering and analysis of information systematically applying concepts and skills to observing phenomena as they appear in the real world is called field study.

Driftwood

"As the traveler who has once been from home is wiser than he who has never left his own doorstep, so a knowledge of one other culture should sharpen our ability to scrutinize more steadily, to appreciate more lovingly, our own."

From *Coming of Age in Samoa* by Margaret Mead

Questions

Write your answers on a separate piece of paper. Use complete sentences.

1. How do the cultures of people affect the way they shape the human landscape?
2. What can the study of the human landscape of your community tell you about the people who moved there from other regions and countries?
3. What is a family tree? How can you use the information from your family tree and the family trees of your classmates to show the distribution and movement of humans?
4. Where can you find the genealogical information that you need to construct a family tree?
5. Where can you find information about the history of your community?

Name

Tracing the History of a Local Family

Purpose

To research the history of a family to illustrate patterns of human migration and cultural diffusion.

Background

A family tree is a graphic display of the generations of family members who have preceded a particular individual or individuals. You may choose to do either your own family tree or the family tree of someone else in the community, such as a friend or other relative. In order to be ready to display and discuss this family tree later during *Rivers Geography,* you will need to start your research as soon as possible. Make sure you allow enough time to interview family members or other people who are knowledgeable about the family. You also will be expected to seek written records, such as a printed family history, to document information about past family members.

Materials

Per student (distributed with this handout)

- base map of your country
- base map of world
- markers—black, blue, red, green

Per student (needed during Lesson 4)

- 15 colored map tacks or straight pins and construction paper squares:
  1 black
  2 blue
  4 green
  8 red
- atlas

Optional (per student)

- list of references about community history
- set of short articles, historic maps, or photographs about community history

Per class

- folders with photocopies of articles, photographs, and other information about community history
- city, regional, nation, and world maps mounted on bulletin boards or similar surface

Procedure

PART A. Making a Family Tree

1. Obtain the names of three generations of family members and their birthplaces (country, state, city), going back to great grandparents. Enter this information on the sheet provided with this activity to create a family tree. If you cannot find out all this information, include notes indicating the steps you took.
2. When you have completed your family tree, plot the birthplaces of the family members on maps of your country and the world. Use a black marker to plot the place where you (or the person whose family you researched) were born. Use a blue marker to plot the birthplaces of parents, a green marker for grandparents, and a red marker for great grandparents.
3. Answer the questions in Observations under "About the Family."

PART B. Learning About Local History

3. Conduct research to find out about your community, using the resources available in the classroom, school library,

public library, and among individuals, as guided by your teacher. Use the information you gather to answer the rest of the questions in the Observations and Analyses and Conclusions sections of this activity. Write on a separate piece of lined paper; use complete sentences.

PART C. Displaying Local Patterns of Immigration

4. When requested by your teacher, pin map tacks (or straight pins with squares of colored paper) on the designated country and world maps to represent the birthplaces of the individuals identified on the family tree you have made. Follow the colors on the legend as indicated by your teacher, such as:

   black pin—you or subject of study
   blue pins—parents
   green pins—grandparents
   red pins—great grandparents
5. Answer the Critical Thinking Questions. On your separate piece of lined paper, use complete sentences.

Observations

A. About the Family

1. When did the family you researched first move to your local community?
2. Where did the family live before they moved to your community?
3. Why did they move to your community?
4. What cultural practices in the homes of this family can be traced back to other parts of the country or world? Describe what you know of their origin.

B. About the Community

5. When was your community first settled?
6. What country or region were the first settlers from? Where were later settlers from?
7. Does your community sponsor celebrations, customs, traditions, or holidays (other than legal holidays) that emphasize the unique character (cultural religious, occupational, or recreational) of the community and the people who live there? Explain.

Analyses and Conclusions

1. Why did people select your site for settlement? Did the location offer advantages for transportation or agriculture?
2. What role did the river, stream, or other waterway play in making the site attractive for settlement? Was the water used for transportation, as a water supply, or as a source of food? Did it help connect the site to other places, or was it a barrier?
3. What role in the community does the river, stream, or other waterway play at present? Does it serve economic, recreational, or other needs?

4. List the human features and activities in your community that would not exist if it were not near the water.
5. Has the river, stream, or waterway been modified to serve human needs? Explain.
6. Has the physical form of the present-day community been influenced by the river, stream, or waterway? What adaptations have been made?
7. Does architecture reveal changes that can be identified with particular periods of time? With particular cultural influences? List examples, including dates of construction and styles of architecture.

Critical Thinking Questions

Answer the following questions by referring to the U.S. and world maps in your classroom on which you and your classmates have plotted the birth-places of people from your family trees.

1. Briefly describe the patterns created by the plotted birthplaces of you and your classmates.
2. What patterns are observable from the plotted birthplaces of the different generations: parents, grandparents, and great grandparents? Are these patterns similar to the pattern for your class? Why or why not?
3. If the birthplaces of more remote generations (earlier than great grand-parents) were plotted on the maps, would you expect the pattern to be similar to or different from the patterns created by the birthplaces of more recent generations? If different, in what ways? Why?
4. If the birthplaces of the next generation (your future children) were plotted, do you suppose the map would assume a similar or different pattern? Why?
5. The concept of **distance decay** states that the frequency of occurrence of a phenomenon tends to decrease as the distance from a central point increases. Explain why people tend to live in or near the place where they were born. What causes people to move away from their birthplaces?
6. How is your community unique? What activities or programs could be established to emphasize its unique character?

Keeping Your Journal

1. Describe any unique cultural or ethnic practices of your community or your family.
2. Select a topic presented in this activity's questions and write a more detailed or more personal entry about it in your journal.
3. Select a particular historic period and write a geographic past for your community.

Family Tree

Great Grandfather	Great Grandfather	Great Grandfather	Great Grandfather
........./........./.................	........./........./.................	........./........./.................	........./........./.................
Country/State/City	Country/State/City	Country/State/City	Country/State/City
Great Grandmother	Great Grandmother	Great Grandmother	Great Grandmother
........./........./.................	........./........./.................	........./........./.................	........./........./.................
Country/State/City	Country/State/City	Country/State/City	Country/State/City
Grandfather	Grandmother	Grandfather	Grandmother
........./........./.................	........./........./.................	........./........./.................	........./........./.................
Country/State/City	Country/State/City	Country/State/City	Country/State/City

Father	Mother
........./........./.................	........./........./.................
Country/State/City	Country/State/City

You or Other Subject of Study

................................./.........../.................................

Country/State/City

Movement and Cultural Diffusion

No location exists in a vacuum, totally isolated from the outside world. Locations are interdependent, and this interdependency gives rise to the movement of people, goods, and ideas from one location to another. These movements show the relationships within and between regions. The study of the geographic theme of movement, as it applies to river locations, will help you see the relationships, or **spatial interactions,** between phenomena.

Movement

Certainly you know that water moves. River water is moved by gravity to the oceans. The water vapor in the atmosphere condenses, then falls as precipitation in another location. Water evaporates and rises as water vapor into the atmosphere. Some of the water that falls as precipitation may be used by plants and then transpired into the atmosphere. Some water collects in rivers or on land and evaporates, returning again to the atmosphere. This is the hydrologic cycle—a system that features movement.

The theme of movement is also illustrated by the daily movements of wildlife and the seasonal **migrations** of species from one location to another. People and goods move by automobiles, airplanes, trains, and water transport. Human communication—another form of movement—occurs by telephone, computer, facsimile, radio, and television. Some movements are less obvious, especially those that involve the deep-seated beliefs of a group of people. Yet cultures do move, and the spread of cultural traits from one area to another is called **cultural diffusion.**

DRIFTWOOD

"Never give children the chance of imagining that anything exists in isolation. Make it plain from the very first that all living is relationship. Show them the relationships in the woods, in the fields, in the ponds and streams, in the village and the country around it."

Aldous Huxley

Cultures

Cultures define who groups of people are. **Cultures** are the unique ways that groups of people speak, dress, worship, make a living, and interact with one another. **Material culture** includes everything in the landscape that human beings have put there, including such features as buildings, transportation and communication lines, dams and levees, farms, factories, mines—even the clothes they wear and the utensils they use to prepare their food. Elements of material culture that have in some way modified the physical environment are called **cultural features.** The material culture of a group is the outward manifestation of **nonmaterial culture**—the ideas, beliefs, customs, and traditions that make one group distinct from another. The distinctive appearance a group's cultural features makes on the area it occupies is called a **cultural landscape.**

When people move, they take their cultures with them. Visible and invisible threads may connect people over great distances and in a variety of ways. Clues to connections can be found in the cultural landscape by observing how people in various communities live and celebrate those events they think important. Customs, traditions, architectural styles, food preferences, religious practices, and other traits may link people and settlements over thousands of miles.

DRIFTWOOD

"America is a great story, and there is a river on every page of it."

From *Life on the Road* by Charles Kuralt

The Influence of Rivers

The availability of fresh water has always influenced where people decide to settle. Because of the advantages to human settlement offered by rivers, it is not surprising that most major cities are located on rivers or on coasts at the mouths of rivers. Indeed, the very earliest urban centers were located in the Tigris-Euphrates, lower Nile, Indus, and Huang river valleys to take advantage of not only supplies of fresh water for domestic use and crop irrigation but also water routes for trade and transportation.

In North America, Native Americans preferred river sites. Some Native American economies were based on fishing. Later, the **diffusion** ("borrowing" or spread) of the use of domesticated plants such as maize (corn), squash, and beans from Mesoamerica (present-day Central America and Mexico) permitted others to establish permanent settlements, also at sites by rivers and streams that supplied fresh water. In Chaco Canyon (in present-day Arizona), the Anasazi had established communities based on irrigated agriculture by about A.D. 1000. In the Mississippi Valley, the large community of Cahokia was supported by agriculture and built around a large ceremonial temple mound.

Early European settlers along the East Coast also selected river sites. The French used the Great Lakes and the St. Lawrence and Mississippi Rivers as travel routes to the continent's interior. In the Southwest, the Spanish established their missions and presidios (forts) at locations where they could obtain fresh water. The forced migration of Africans focused primarily on the continent's southeast, where plantation settlements depended on water transportation for their agricultural exports. As more **immigrants** (IHM ih gruhnts) arrived in North America and expanded existing settlements, people used rivers as routes to new locations. River cities are "breaks" in transportation—points where goods must be transferred from one type of carrier to another, a task that provides job opportunities. At other river sites, early factories were established to take advantage of rapidly moving water as a source of power. Later, a series of canals linked the natural waterways. For instance, the Erie Canal connected the Hudson River with the Great Lakes, spurring new settlers into the interior and establishing New York as the East Coast's dominant city.

Human Migration and Cultural Interaction

Emigration (em uh GRAY shuhn) is the process of leaving one's native country to settle in another. Geographers refer to the reasons why people move as **push factors**—attempts to escape undesirable conditions—or **pull factors,** desires to move because of real or perceived opportunities elsewhere. The promise of free or inexpensive land, escape from religious persecution, political instability or oppression, and economic misfortune all influenced movements to North America. In some cases movement was not voluntary, as in the removal of Native Americans from their land and the transport of Africans to North America for slave labor. European countries provided most of North America's early immigrants. In recent years, many Asian and Central American people have migrated to North America. By the twentieth century, the large cities that had grown up along coasts and rivers had become the destinations of the new immigrants.

The process by which people come into a new country or area is called **immigration** (ihm uh GRAY shuhn). Each **ethnic group** that has come to North America has brought its own unique culture. As cultures interact, they change. When each ethnic group contributes something to the overall culture, the result is referred to as a **melting pot.** A **mosaic** of cultures will result if each ethnic group retains its cultural identity. In either case, the largest ethnic group tends to have the greatest influence on the overall culture.

Cultural pluralism occurs when more than one culture exists at the same time in the same region. Small cultural groups that are distinct from the larger culture are termed **subcultures** and are often associated with particular locations called **culture areas** or **culture islands.**

Cultural pluralism exists in both Canada and the United States. Although English is the majority language in Canada, a French Canadian culture area is centered in Quebec. Native American culture areas also exist.

Many distinctive culture areas are observable in the United States, particularly in large cities. For example, New York City has become a culture area for Puerto Ricans who have moved to the mainland. South Florida is a culture area for Cubans and immigrants from other Caribbean islands. Mexican and Central American immigrants are found in large numbers in Southwest states bordering Mexico. East Asians have settled in largest numbers in the West Coast states. African Americans, long separated by residential areas, constitute a subculture. A retirement subculture continues to grow, especially in the "Sun Belt" states of the South.

Questions

Write your answers on a separate piece of paper. Use complete sentences.

1. How does the geographical theme of movement apply to people, goods, and ideas?
2. What are the differences among material culture, nonmaterial culture, and cultural landscape?
3. How have rivers influenced the movement of people?
4. Why do people emigrate?
5. What happens when different cultures interact?

TEACHER NOTES

How Do Humans Interact With the River?

Focus: Human-Environmental Interaction

In this lesson, students will explore why, how, and with what impacts humans have modified rivers and streams. They will experience the complexity of environmental decisionmaking, including recognizing that decisions-makers must consider many different interests and points of view in order to avoid simply passing problems on to someone else.

Perspectives and Geography Standards

Perspectives: Ecological and Spatial

Geography Standards Receiving Primary Emphasis

This lesson primarily promotes mastery of the following standards:

13. How the forces of cooperation and conflict among people influence the division and control of Earth's surface
14. How human actions modify the physical environment
18. How to apply geography to interpret the present and plan for the future

Geography Standards Receiving Secondary Emphasis

All other standards receive some secondary emphasis.

Learner Outcomes

Students will:

1. Learn how humans have modified rivers and streams.
2. Discover how people make decisions about the environment.
3. Role-play and interact with others to solve hypothetical environmental problems, then analyze and evaluate the role-play.
4. Prepare a written environmental impact statement.

Time

Four class periods of 40–50 minutes per period

BEFORE STARTING: Assign Student Information 5.1: Human Modifications of the River as homework

DAY 1: Student Information 5.1: Human Modifications of the River
Student Information 5.2: Making Decisions About the Environment
Set up Student Activity 5.3: Simulation of Environmental Decisionmaking (use corresponding Teacher Guide)

DAY 2: Student Activity 5.3: Simulation of Environmental Decisionmaking, Part A

DAY 3: Student Activity 5.3: Simulation of Environmental Decisionmaking, Part B

DAY 4: Student Activity 5.3: Simulation of Environmental Decisionmaking, Part C

Advance Preparation

Prepare to distribute to each student a copy of Student Information 5.1 and 5.2 and Student Activity 5.3. To prepare materials for Student Activity 5.3, duplicate all the role-play materials included on pages 134–143 in corresponding Teacher Guide. Photocopy one regional map for each person in the corresponding group. Laminate one master regional map per group. Cut out and laminate regional descriptions, regional problems, and role cards separately. Rubberband each region's role cards together. Into a large envelope labeled "Region A" place the laminated map and description, and photocopied maps for one region. Repeat for Regions B, C, and D. Into a large envelope labeled "Role Cards" place all four groups of role cards. Into a final envelope labeled "Problems" place all four problems.

Assign Student Information 5.1: Human Modifications of the River as homework the day before starting Lesson 5.

Materials

Student Activity 5.3: Simulation of Environmental Decisionmaking

For the teacher, to prepare packets for student groups

6 large envelopes
4 laminated regional descriptions (masters in corresponding Teacher Guide)
4 laminated regional maps (masters in corresponding Teacher Guide)
1 photocopy of the group's regional map per group member
4 sets of laminated role cards (masters in corresponding Teacher Guide)
4 rubber bands
4 laminated regional problems (master in corresponding Teacher Guide)

For the teacher

1 photocopy of each regional map
masking tape

Vocabulary

conservation
environmental impact study
environmental modification
NIMBY (Not In My BackYard)
recycling
region
role playing
sanitary landfill
simulation
special-interest group
waste

Background for the Teacher

The role-playing in Student Activity 5.3 is designed to simulate conditions within which individuals must operate in order to solve real environmental problems involving rivers and streams. It emphasizes opportunities for participants to share ideas and enhance their decisionmaking skills. Activities such as these, which involve students directly, usually generate significant enthusiasm.

Each student becomes a member of a community facing a problem involving the river. Students become familiar with their individual roles and their community before considering the problem. As they focus on the decision ahead, a "community" meeting forces them to consider points of view in conflict with their own.

A spokesperson for each group reports to the class his or her group's problem, conflicts, and decision. Only after all groups have made reports does the class discover that all four regions are part of the same watershed and that the impact of each decision will be felt by others. For students, the problem then shifts from a local issue to a larger, regional one. They must then consider the impact of a decision on not only themselves and their own communities but the region of which they are a part.

Introducing the Lesson

1. Give students a brief overview of Lesson 5. Discuss Student Information 5.1: Human Modifications of the River. (Sample answers are in Appendix B.) Ask for examples of the following major points as they apply to your local river or stream and community:
   - Physical environments pose challenges for human use.
   - Humans modify the physical environment to serve their needs.
   - Such modifications may depend on perceived needs, cultural beliefs, and available technology.
   - Interactions may be simple or complex and on large or small scales.
   - Well-intentioned changes may produce negative results.
   - Controversy may surround proposed changes.
   - Changes that serve human needs may be detrimental to natural systems.
2. Have students read and answer questions in Student Information 5.2: Making Decisions About the Environment.
3. Review Student Information 5.2 and discuss student answers. (Sample answers in Appendix B.)

Developing the Lesson

1. Divide the class into four groups, and have them carry out Part A of Student Activity 5.3: Simulation of Environmental Decisionmaking. (Use corresponding Teacher Guide.) (In classes of 16 or fewer students, two groups of eight may stage two simulations, in turn, of two different regions. In classes of eight or fewer students, students can simulate all four regions on successive days. The first five roles in each group *must* be played.)
2. When each group has arrived at a decision regarding their land-use problem and clarified how they came to that decision, have the class do Part B of Student Activity 5.3.
3. Present the problem for Part C of Activity 5.3 (see corresponding Teacher Guide). Provide time for students to work individually or in their groups to complete the Analyses and Conclusions segment.

Concluding the Lesson

1. Reassemble students into regional groups. Have each group complete Part C, a written environmental impact report describing the expected impact of the proposed project on the entire region.
2. Use group presentations of their reports as prompts for further discussion.

Assessing the Lesson

1. As appropriate, evaluate student participation in the group process for Student Activity 5.3.
2. As appropriate, collect and assess Student Activity 5.3 answers and group reports. Here is a suggested scoring rubric, which you can use with the performance criteria included in the student activity sheet:

Scoring Rubric

Score	Expectations
0	No response attempted.
1	Demonstrates only rudimentary understanding of changes caused by the proposed project; may not address all performance criteria.
2	Demonstrates satisfactory level of understanding of changes. All performance criteria are addressed, although explanations may be incomplete.
3	Demonstrates excellent level of understanding of changes. All performance criteria are explained. Explanations demonstrate an understanding of cause and effect and interrelationships of elements.

Extending the Lesson

1. Students may construct a land-use map as a way to evaluate the human impact on the study region environment. Working in teams, students create a base map by drawing enlarged sections of the study-area map at the same scale or use a laminated topographic map(s). Have students include such land-use categories as agricultural, mining, industrial, residential, commercial, public and semipublic (parks, schools, hospitals, churches, cemeteries), adding others as appropriate.

   Encourage students to use colored markers to superimpose on the natural and cultural features of the topographic map their personal knowledge of the region as well as information from such sources as plat books, county or regional planning and zoning boards, and conservation departments. Direct them to fill in the map with color shadings corresponding with each type of land use. To indicate that two land uses (for example, residential and commercial areas) occupy the same space, they may use alternating diagonal lines of appropriate colors.
2. As an alternative to playing the roles provided for Student Activity 5.3, assign student groups to identify and write descriptions for typical roles for their assigned regions. They should identify the same number of roles as the group has members. Students can then draw roles by lot before the group receives its regional problem.
3. Have students examine the process of problem solving directly. Have them cite which parts of Student Activity 5.3 illustrate each step of the process, such as defining problems, hypothesizing, gathering information or doing research, and stating conclusions or making generalizations.

GUIDE

For Student Activity ***5.3***

Simulation of Environmental Decisionmaking

Purpose

In parallel with Student Activity 5.3: Simulation of Environmental Decisionmaking, to provide specific teaching steps that facilitate students role-playing decisions about land and water use. (Suggested answers are in italics; notes for teacher are in bold.)

Materials

Per class
4 large envelopes, labeled "Region A," "Region B," "Region C," and "Region D;" each containing laminated regional description, laminated regional map, photocopy per group member of regional map
large envelope containing 4 rubber-banded groups of laminated role cards
large envelope containing 4 laminated regional problems
For the teacher
1 photocopy of each regional map
masking tape

Background

This Teacher Guide provides teacher notes and sample answers for Student Activity 5.3. For instructions on how to prepare the envelopes and laminated materials, see Lesson 5 Teacher Notes, Advance Preparation.

Student Activity 5.3 is divided into three parts. Students should read and complete one part before reading the next part. In Part A, students become acquainted with their hypothetical region and community, the problem faced by the community, and the roles they will play. In Part B, each group simulates a community meeting in which individuals present different points of view and the group makes a decision about how to resolve the region's problem. In Part C, students learn what decisions the other three communities made and how these decisions impact the river system as a whole. Then each group prepares an environmental impact statement.

PART A. Simulation of Community Decisionmaking

Procedure

1. Join a specified group as instructed by your teacher. **Divide the class into four groups. Arrange students in each group so that each can see and hear all group members, groups are separate from one another, and you can circulate among the groups.**

2. When your teacher distributes to your group a map and a description of your hypothetical region, examine these materials and answer the first question in the Observations section.

Distribute Student Activity 5.3. To each group, provide the large envelope with regional description and maps. Explain Student Activity 5.3 to students, using Background as appropriate. When students have reviewed description and maps, distribute role cards.

3. When your teacher distributes the cards that describe the roles that each member of your group will play, choose roles as directed by your teacher.
4. Playing your role, work with your group to complete the observations, analyses and conclusions that follow, written or orally, as directed by your teacher.

Observations

1. Locate the features of your region on the map. Briefly describe your community and region. *Responses should correspond to the materials distributed to each group.* **One student in each group may read the regional description aloud.**
2. Read your role and write a description of it.
3. Have each person in your group read aloud his or her role. In your own notes, record descriptions of each role.

2–3. *Responses should correspond to the materials distributed to each group.*

4. Read the problem your community faces, and write it in your own words. Responses should correspond to the materials distributed to each group. Distribute to each group its land-use problem. Suggest that one student read the problem aloud to the group. Have students read their problem summaries aloud.

Analyses and Conclusions

1. Based on your role, write down two solutions to the problem that you would consider acceptable and one that you would consider unacceptable. *Each student should identify at least one acceptable site. In Region D, the only acceptable site is outside the region. For some roles, more than one site is acceptable; other roles clearly indicate sites that are unacceptable.*
2. Briefly note three arguments that could be used to win others over to your point of view. *Answers will vary with the role played. Students should argue against sites that would have a negative impact on them and argue for sites that have the fewest drawbacks or have a positive impact on them.*
3. As a group, hold a "community meeting" to discuss the solutions and supporting arguments suggested by your group. As others voice their opinions, place an "X" on your map for each objection to the use of that site to solve the problem. As a group, decide on a solution that everyone in your group finds acceptable. Write the solution your group finally agreed on.

Regional groups A, B, and C usually select site C, as it would have the least negative impact on the role players. Depending on the effectiveness of the Real Estate Broker player in Region D, arrangements should have been made to purchase land in Region A.

When groups hold their "community meetings," circulate among the groups to encourage participation. In particular, you will need to encourage the Real Estate Broker for Region D to seek land for purchase outside of Region D. (After a few moments of discussion, students will discover that no one in Region D is willing to sell land. Meanwhile, the solution in Region A becomes easier if the Farmer is able to sell land.). Impress upon students that each must play his or her role in the community meeting and take a stand on the problem. Set a time limit for deliberation.

4. How did your group reach a resolution to the problem? What arguments were most convincing? *Resolution of the problem was probably made by a show of hands. Those students who "got into" their roles probably put forth the most strenuous arguments, although not necessarily in a successful effort.*

PART B. Simulation of Regional Decisionmaking

Procedure

Have students read Part B.

1. Select a member of your group to briefly describe your region and community and the roles played, and to present your community's problems addressed, choices considered, and the group's solution. **While each presentation is made, display the map of the corresponding region.**
2. As you listen to the presentations, record each group's regional description, problems and solutions. *Answers may be sketchy, as they will be based on notes taken during the oral reports of other students.* **You may want to note on the chalkboard the problem and solution for each region. After each presentation, allow time for questions.**
3. Participate in a regional meeting in order to complete the analyses and conclusions that follow.

Analyses and Conclusions

Fold each map sheet at top border; then tape the four maps together on the chalkboard or wall to reveal that all are parts of the same river system. (See figure on next page. Map A goes at top, with Map B below it, then Map C, then Map D.) In most cases, each group's decision will cause a problem for another group. Permit time for students to express their reactions to the conflicts. Have students complete the Analyses and Conclusions questions (perhaps as homework).

1. Considering the overall impact of these solutions, what was the net result of the four decisions? *The "solutions" simply passed the problems on to other locations, but the problems remained within the watershed region.*

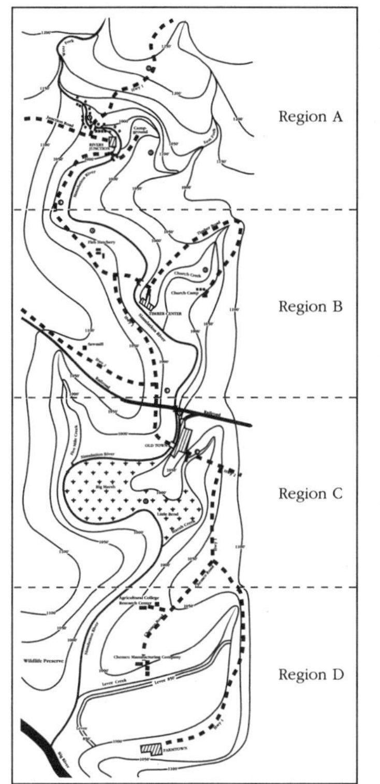

Combined Regional Maps

2. How does the impact of the combined decision differ from the decisions made by the individual regions? How can problems solved in one region affect other regions? *On a local scale, a problem can be eliminated by changing its location, but on the larger scale, the problem remains. Students are likely to reevaluate the concept of waste. They may conclude that the term needs a new definition and that waste disposal requires more than a change of location. Problems of one region should be seen as problems for all regions because all regions are connected.*
3. How can such problems be resolved to the satisfaction of all? *Decisions about a watershed must consider all interests, rather than just some. Waste must be reduced. Waste that remains should be recycled or disposed of in such a manner that it does not pose a threat to people or ecosystems.*
4. Participate in a "regional meeting" in which the interests of each group and the conflicts between different groups are discussed. As you hold this discussion, refer to a map that shows all the regions. Try to arrive at a solution that includes all four regions. **Lead a "regional meeting" in which students discuss the interests of each group and the conflicts between different groups. As appropriate, refer to the map that combines the four regions into one watershed region. Encourage students to arrive at a solution that includes all regions.**

PART C. Environmental Impact Study

Procedure

Reassemble students in regional groups. Pose the following situation:

A government-financed plan has just been released. It will cause profound changes in the river region. A lock and dam will be constructed near the mouth of the river and will create a pool at an elevation of 300 meters (1000 feet). The plan cites three reasons for the project: to provide flood control, to promote water-related recreation in an economically depressed region, and to provide a channel for barge traffic to handle large shipments of a very valuable mineral resource recently discovered between East and West Forks.

The announcement may provoke a variety of student responses, depending on how intensely students identify with the roles they have played. Have students read Part C.

1. Listen to the proposed change described by your teacher. Take notes as needed.
2. Respond to the analyses and conclusions that follow. **Provide time for students to work individually or within their groups to complete the questions. As appropriate, facilitate additional discussion within groups and as a class.**
3. Using those answers, work with your group to prepare an environmental impact report that predicts the expected effects of the proposed human alteration to the natural environment. **As needed, review the nature of an environmental impact study.**

Analyses and Conclusions

1. List as many changes as you can that the proposed construction is likely to cause in the physical landscape. Include environmental changes and human modifications that are likely to occur as people adapt to their new environment.*(a) The floodplain will be covered with water; creeks will be reduced to their headwaters; the marshes near Old Town will be covered. Consequently, a large part of the wetland ecosystem will be destroyed. (b) The free-flowing streams will become part of a large, artificial lake. Some aquatic life forms will be unable to adapt to the change. (c) Vegetation on land at elevations less than 300 meters (1000 feet) above sea level would be flooded. (d) It is likely that a deeper channel would be dredged through the marsh area near Old Town to facilitate barge traffic.*
2. List as many changes as you can that the proposed construction is likely to cause in the human environment. *a) The low population density of the region means that few people would be displaced by the new lake. Nevertheless, some houses would be affected. Other human features that would be affected include bridges, roads, the campground near Rivers Junction, Church Camp near Timber Center, part of Chemco Manufacturing Company, farmland near Farmtown, and parts of the communities of Rivers Junction and Timber Center. (b) Activities that would be disrupted include farming and chemical manufacturing near Farmtown, commercial and recreational hunting and fishing in the marshes near Old Town, religious conferences at Church Camp, and recreational boating and camping at Rivers Junction. (c) Communities would be less accessible to one another because of the disruption of highway routes.*
3. How will humans adapt to the new environment? *(a) A government agency may prevent or control development of human activities on the shoreline. (b) If individuals are able to make decisions about the shoreline development, it is likely that new types of land uses will be developed, such as residential construction, businesses to cater to visitors (hotels, restaurants, amusements), and water recreation. (c) The farmers around Farmtown will probably decide to raise the levee that protects their farms. (d) The decisions made initially to solve each community's problem will need rethinking. The proposed lake and mining operations impact all these decisions.*

Creating an Environmental Impact Report

1. Based on your observations, list the advantages of the proposed construction. *Advantages may include any of the following: (a) A valuable mineral resource can be obtained. (b) Opportunities for new types of businesses and recreation will be developed. (c) The proposed lake may encourage the development of tourism in the region. (d) The dam will permit the water level to be held at a more consistent level. (e) The project may lead to the improvement of transportation routes.*

2. List the disadvantages of the proposed construction. Remember, a particular change may be considered beneficial from one point of view and nonbeneficial from another. Try to view possible changes from a variety of perspectives. *Disadvantages may include any of the following: (a) People, homes, transportation routes, businesses, and industries must be removed or relocated. (b) Jobs will be lost. (c) Wetlands and wetland ecosystems will be eliminated. (d) A free-flowing river will become a lake. (e) Such a project is expensive to construct. (f) Some forms of recreation will no longer be possible.*

   Before having students do Question 3, reassemble them into regional groups and review the question. Emphasize that each group should prepare one report. Students should use answers to the Part C questions to prepare their reports.
3. Together with other members of your group, prepare a written environmental impact report that lists and explains the major impacts the project will have on the physical and human environments of the watershed. Though you cannot anticipate all effects, you can predict many with accuracy. *Reports will probably be restatements of the answers to questions in this section, although discussions may raise other concerns.*
4. Select a member of your group to present your report to the class. Be prepared to assist by offering explanations in the discussion that will follow. **These readings should prompt further discussion. Encourage students to consider various perspectives before they conclude that a change is good or bad. Open the discussion to all members of the class. Ask them to consider the impact of the changes on the persons whose roles they portrayed in the simulation. Propose hypothetical changes in the physical environment of the local community, and ask students to consider the advantages and disadvantages of each change.**

   As appropriate, have students complete the Keeping Your Journal section.

Scoring Rubric

Score	**Expectations**
0	No response attempted.
1	Demonstrates only rudimentary understanding of changes caused by the proposed project; may not address all performance criteria.
2	Demonstrates satisfactory level of understanding of changes. All performance criteria are addressed, although explanations may be incomplete.
3	Demonstrates excellent understanding of changes. All performance criteria are met, and explanations demonstrate an understanding of cause and effects and interrelationships of elements.

Region A Description, Problem, and Roles

Region A Description

The small community of Rivers Junction (pop. 150) serves as the outfitter for canoeing and camping in this heavily wooded, hilly region. Summer homes and cabins offer "get-away-from-it-all" retreats for people from outside the region, who are drawn to the beautiful woods, hills, and clear, fast-moving streams.

Residents are employed in activities that largely cater to seasonal visitors. Annual income is low, and most people can be classified as underemployed.

Cattle raising is of some importance to the region. Uneven terrain makes crop farming difficult.

Because of its small size, the community has no waste treatment plant. Most dwellings are served by individual septic systems, many of them antiquated.

Region A Problem

Concerned citizens gather to discuss with a farmer of the region the farmer's plans to build a large feedlot—a plot of land on which livestock are fattened for market. The farmer reveals to the citizens that sites A, B, and C on the map are under consideration as locations for the feedlot.

Region A Roles

Role Card 1A Summer Home Owner: Rivers Junction is a great place for you and your family to spend vacations and weekends. You love the small town atmosphere and outdoor activities, especially fishing and canoeing. Most part-time residents' cabins front on West Fork, just upriver from Rivers Junction. The community's businesses depend on people like you for survival. Sometimes you are disturbed by the large number of canoeists and campers who interrupt the peace and quiet of the town.

Role Card 2A Canoe Outfitter: Canoeing is very popular in the Rivers Junction Region, attracting many visitors from outside the region during the busy summer months. East Fork is most important to your business because of the scenic beauty of occasional rapids. On West Fork, too many summer cabins have sprung up, destroying the scenery. Without your business, you would be forced to find work outside the region, because few other jobs exist locally.

Role Card 3A Restaurant Owner: Rivers Junction has only one restaurant. You must depend on local residents, part-time residents, and outside visitors to stay in business.

Role Card 4A Farmer: You own a considerable amount of land in the Rivers Junction Region, although much of it is wooded and of little use for raising cattle, which is your primary activity. You have been toying with the idea of starting a large feedlot to fatten cattle for market, believing it would be profitable for you and good for the economy of Rivers Junction, which now has to exist on handouts from Big City visitors. Of the land you own, sites A and B are preferable to C, which would cost more to develop. If you could find a buyer of some otherwise worthless land (site D), you could finance site C.

Role Card 5A Student: You are a high school senior, soon to graduate. You like Rivers Junction and would like to remain, but job prospects are very slim in the region.

Role Card 6A Campground Owner-Operator: Camping is very popular in the Rivers Junction region, and your campground receives high marks from campers. They like the small town, the clean water, and the deep forests.

Role Card 7A Commuter: Although you drive to a job outside the region, you choose to live in Rivers Junction because you enjoy the small town and believe it is a good place to raise your children. You like the place the way it is and hope the tourist trade will not spoil it. You would be willing to pay higher taxes to support the services the community lacks.

Role Card 8A Retiree-Unofficial Mayor: Having lived in Rivers Junction all your life, you have a great desire to see the community prosper. You are pleased that it has become attractive to visitors from outside the region. You would like to see the town grow.

Region B Description, Problem, and Roles

Region B Description

The region around Timber Center (pop. 400) is quite rugged and largely covered by timber. The timber industry provides most of the jobs in this region—logging, hauling timber, and working in the sawmill across the river from the community. Because of increased labor and transportation costs, the mill has had to cut operating expenditures and has laid off half its employees.

Only a handful of owner-operated businesses exist in town, providing basic goods and services. A church camp and fish hatchery operate within the region, but they have few ties with the community other than employing some residents.

Region B Problem

Concerned residents are meeting with a paper mill company representative to discuss the company's plans to build a paper mill somewhere in the region. The company representative reveals that sites A, B, and C on the map are under consideration as possible locations for the new paper mill.

Region B Roles

Role Card 1B Sawmill Worker: Despite 20 years on the job at the sawmill, you have recently been laid off. New job prospects in the region are scarce, and you have few other job skills.

Role Card 2B Church Camp Director: As director of an interdenominational church camp, you arrange and host meetings and conventions the year round. You take great pride in maintaining the facilities and grounds, which front on Church Creek and are surrounded by deep woods. It is the natural beauty that continues to draw groups to the camp.

Role Card 3B Mayor: Your town is facing severe economic problems, even more so now that the sawmill has laid off half of its labor force. With very little to tax (Church Camp is tax-exempt), your community cannot provide adequate basic services. New businesses would increase tax revenues.

Role Card 4B Paper Mill Company Representative: Your task is to select one of three sites (owned by the company you represent) for the construction of a new paper mill. All the necessary resources—large forests, a plentiful water supply, and a cheap labor source—are available at each site.

Role Card 5B Director of Fish Hatchery: The quality of the river water is very important to your job, which is to hatch and release fish into the river and its tributaries.

Role Card 6B Boating and Fishing Enthusiast: You know the river better than most other residents of the region because you spend most of your time there. Your small income, derived from the fish you catch and sell, is supplemented by trapping in wintertime and performing odd jobs.

Role Card 7B Owner of Auto Service and Repair Shop: You barely make a living at your job. Your opinion is that the community needs an economic shot in the arm. More people with jobs, and therefore more money to spend, would help your ailing business to survive.

Role Card 8B Manufacturer's Representative-Naturalist: Your job as a manufacturer's representative requires you to travel outside the region. When you are home, however, you spend much time hiking and exploring the forests around your community. Despite the importance of the timber industry to the local people, you would like to see the forests preserved, perhaps as a park.

Regional C Description, Problem, and Roles

Region C Description

Old Town (pop. 20,000) was once a thriving manufacturing center, but the abandonment of the railroad line has left many former businesses and industries vacant. Some well-to-do families remain. Many others drive to distant communities for jobs. Unemployment is high and the number of people who live in or near the poverty level is also high.

A bond referendum to update the community's antiquated waste treatment plant was overwhelmingly defeated because it would require an increase in taxes. At present, wastewater is only partially treated, and frequent flooding causes the system to overflow and dump raw sewage into the river.

The community's most significant economic activities are found in the old historic district near the waterfront. Antique shops and restaurants draw visitors from outside the region.

Big Marsh is a popular fishing and waterfowl hunting area. In recent years, fish and waterfowl have declined, because of the increased pollution of the river. Most of the pollution is from the waste treatment plant and sources upstream.

Region C Problem

Old Town's city council has just learned that state funds have been obtained to build a new waste treatment plant. Concerned citizens gather at the council meeting, where councilors will decide whether to build the treatment plant at site A, B, or C.

Region C Roles

Role Card 1C Owner-Operator of Marina and Sporting Goods Shop: Your store provides for the needs of people involved with commercial fishing, recreational fishing, and waterfowl hunting. Big Marsh is your meal ticket.

Role Card 2C Social Security Pensioner: Even though your income is quite low, you are managing to get by now that you have moved to government-subsidized housing at the east end of town near Marsh Creek. One of your pleasures is to take your daily nature walk along the footpath beside the creek.

Role Card 3C Chamber of Commerce Representative: Your task is to promote the few businesses Old Town has left, particularly the riverfront area's antique shops and restaurants. The community has spent a great deal of time, effort, and money to "spruce up" the area to make it appealing to customers.

Role Card 4C Commercial Fishing Organization Representative: Commercial fishing, which provides needed jobs and income for your community, depends on the well-being of the river. The best area is south of Old Town at Little Bend before Marsh Creek joins the river.

Role Card 5C City Council Member: As an elected representative, you attempt to do what is best for most people. One of your priorities is to promote and provide basic community services.

Role Card 6C City Engineer: You are consulted on all city building plans. Your advice is to build new facilities near the established city to enhance the convenience of city services and to avoid construction on the frequently flooded lower areas.

Role Card 7C Construction Worker: Most jobs you work on are outside the community, sometimes requiring you to stay away from home for several days at a time. You would like to see more building projects in town so you could work close to home. You supplement your income by doing some boat motor repair, and your spouse works in an antique shop downtown.

Role Card 8C Factory Worker: When the city was prosperous, you had a well-paying job. That job was lost when industries moved away, and you were forced to take a lower-paying job at another factory. You blame the city government for not doing more to keep the factories and the railroad. You are also bitter because much city tax money has been spent to promote the antique shops and restaurants downtown.

Region D Description, Problem, and Roles

Region D Description

Prosperous farming characterizes the broad floodplain surrounding the community of Farm Town (pop. 1,000), which is populated by the close-knit descendants of families who have occupied the same land for many generations. A levee, pumping stations, and an extensive irrigation network are supported financially by a cooperative ownership of landowners and permit high yields of vegetables on the intensively cultivated floodplain.

The people are very proud of their shared ethnic heritage and prosperous lifestyle. An annual festival with tours of historic sites and farms and ethnic foods, costumes, dances, and entertainment draws large numbers of visitors from outside the region.

A source of irritation to the farming community is the chemical manufacturing plant across the river. The plant produces offensive odors and dumps waste into the river. The farmers are especially concerned by the waste in the river, which is their source of irrigation water for crops. On the other hand, the taxes paid by the plant help provide community and regional services. Furthermore, the chemical plant employs many local residents.

Region D Problem

Concerned residents of the region confront the plant's chemical engineer about accidental chemical spills into the river and rumors of improper storage of hazardous wastes on the property. The engineer reveals that the company is trying to purchase land on which to temporarily store the toxic waste until it can be disposed of permanently. The plant engineer also states that the company has engaged a real estate broker to find land the firm can purchase.

Region D Roles

Role Card 1D Floodplain Farmer: Your ancestor first settled the land on which you live and make a living, and you intend to protect it at all costs. You believe the chemical company across the river is polluting the water that you and others use for irrigation.

Role Card 2D Wildlife Preserve Officer: The wildlife preserve you manage is your whole life, and environmental misuse greatly concerns you. You are uncomfortable with all your neighbors: the chemical company, which you believe is improperly disposing of dangerous wastes; the farmers, who you believe overuse insecticides and pesticides; and the agricultural research center, because it occupies land that you would like to see added to the wildlife preserve.

Role Card 3D Chemical Plant Engineer: Your job is to make sure manufacturing wastes are disposed of properly. You are occasionally angered by communities and individuals who allow raw sewage, agricultural runoff, and other wastes to be carried away by the river and yet criticize you when accidental spills occur at the chemical plant. You believe you do your job well, but you now face a very important task: finding a temporary place to store barrels of toxic waste until they can be permanently and properly stored. No space exists on the plant property, and you must look outside the region to buy land. You hire a real estate broker to help you.

Role Card 4D Real Estate Broker: Because the land around Farm Town is seldom sold, you are very interested in the request by the chemical plant engineer to buy land. Your instructions are to purchase a parcel of isolated land, outside your region if necessary. You may not reveal either who the prospective buyer is or how the buyer intends to use the land. The chemical company desperately needs the land and will pay whatever price is necessary to secure it.

Role Card 5D Director of the Agriculture College Research Center: Although your work does not depend on the floodplain farmers across Levee Creek and you see them as somewhat distrustful of outsiders, you have more in common with them than either the chemical plant people or the director of the wildlife preserve. Both would like to control the land occupied by the research center, but because of your dedication, the college is not likely to sell the land at any price.

Role Card 6D Chemical Labor Union Representative: Your interests are in maintaining jobs and insuring the safety of workers at the chemical plant.

Role Card 7D Trucker: Your interests are closely tied to the farmers of the region, because they pay you to haul materials for their operations. Their concerns are therefore your concerns. Any disruption of farming activity would disrupt your business as well.

Role Card 8D Director of Community Festival: Your primary avocation is to direct the annual community festival, which is a very popular attraction for both visitors and residents. The general well-being of the community is important to the festival's success, and you spend a great deal of time on beautification projects and other community programs.

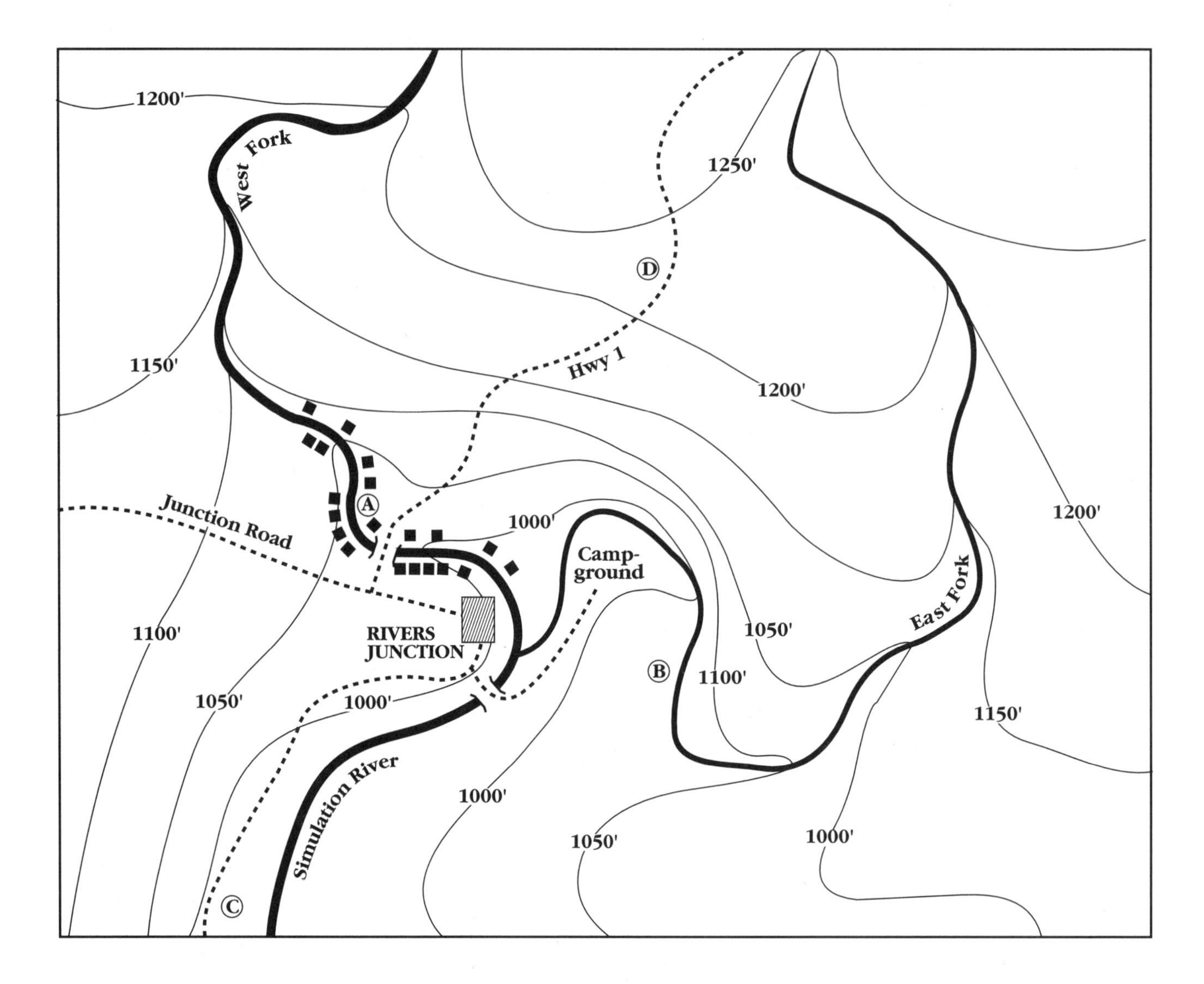

Scale: 1 inch = 1 mile

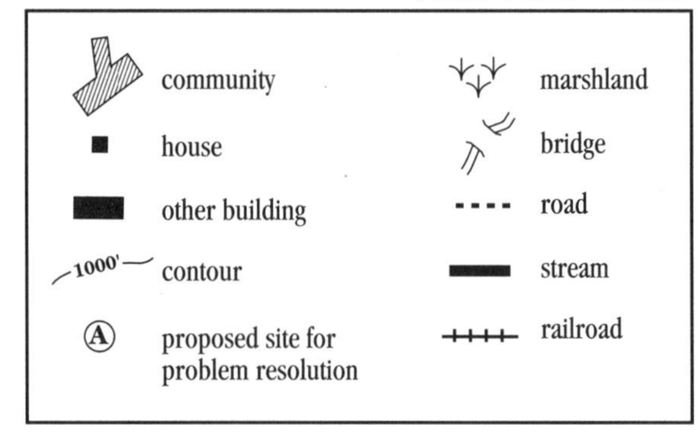

MAP OF REGION A

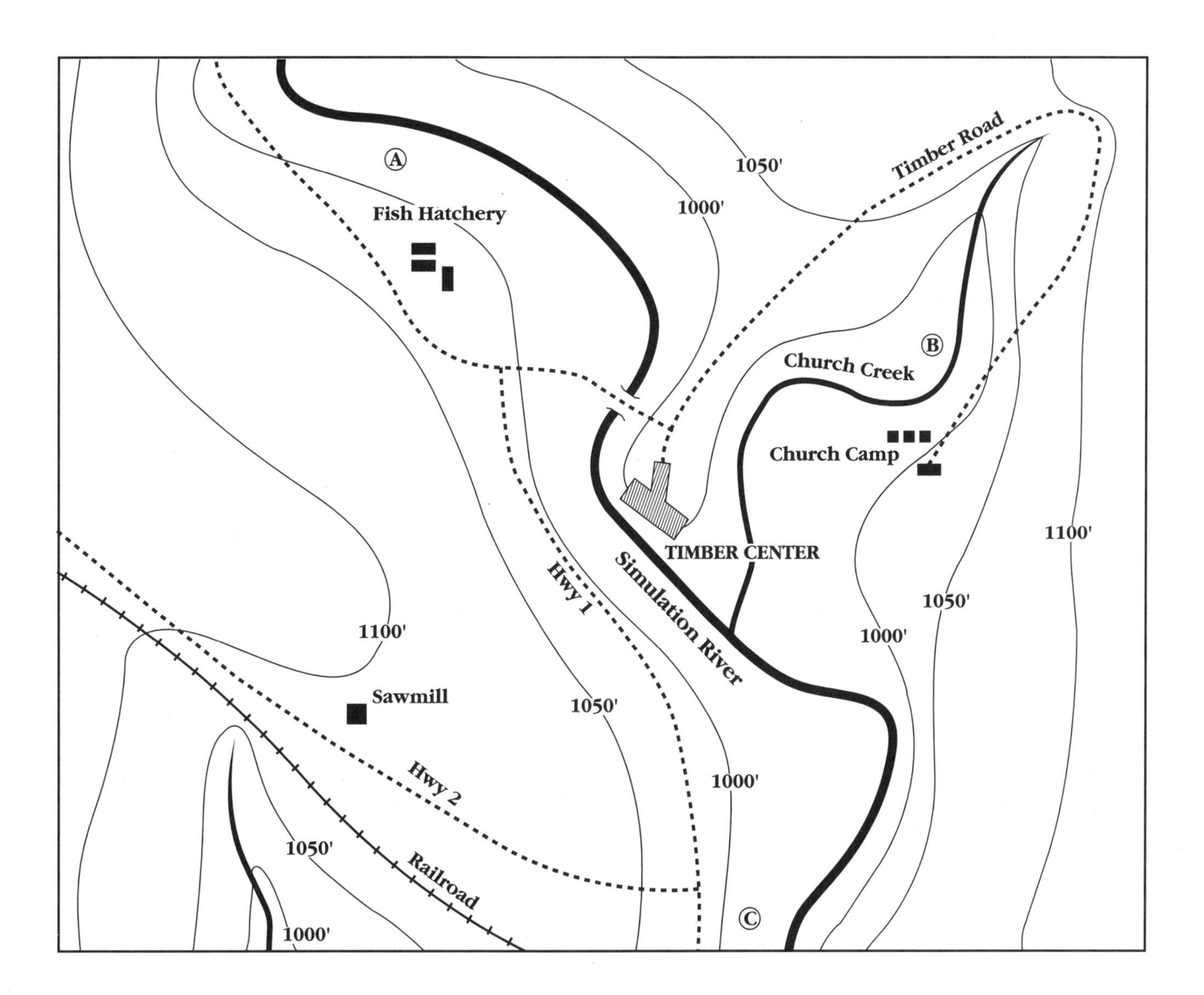

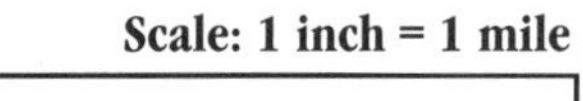

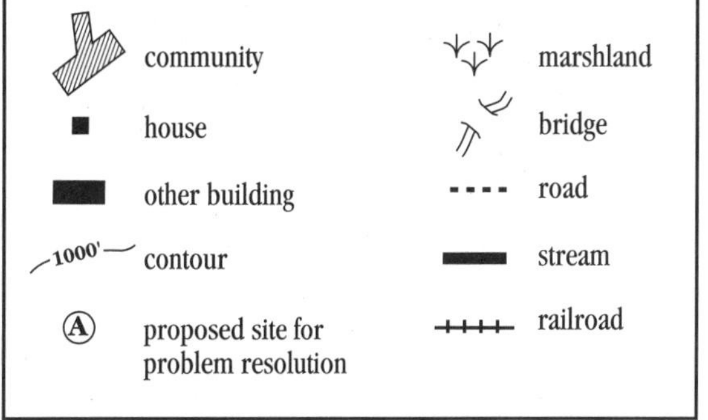

MAP OF REGION B

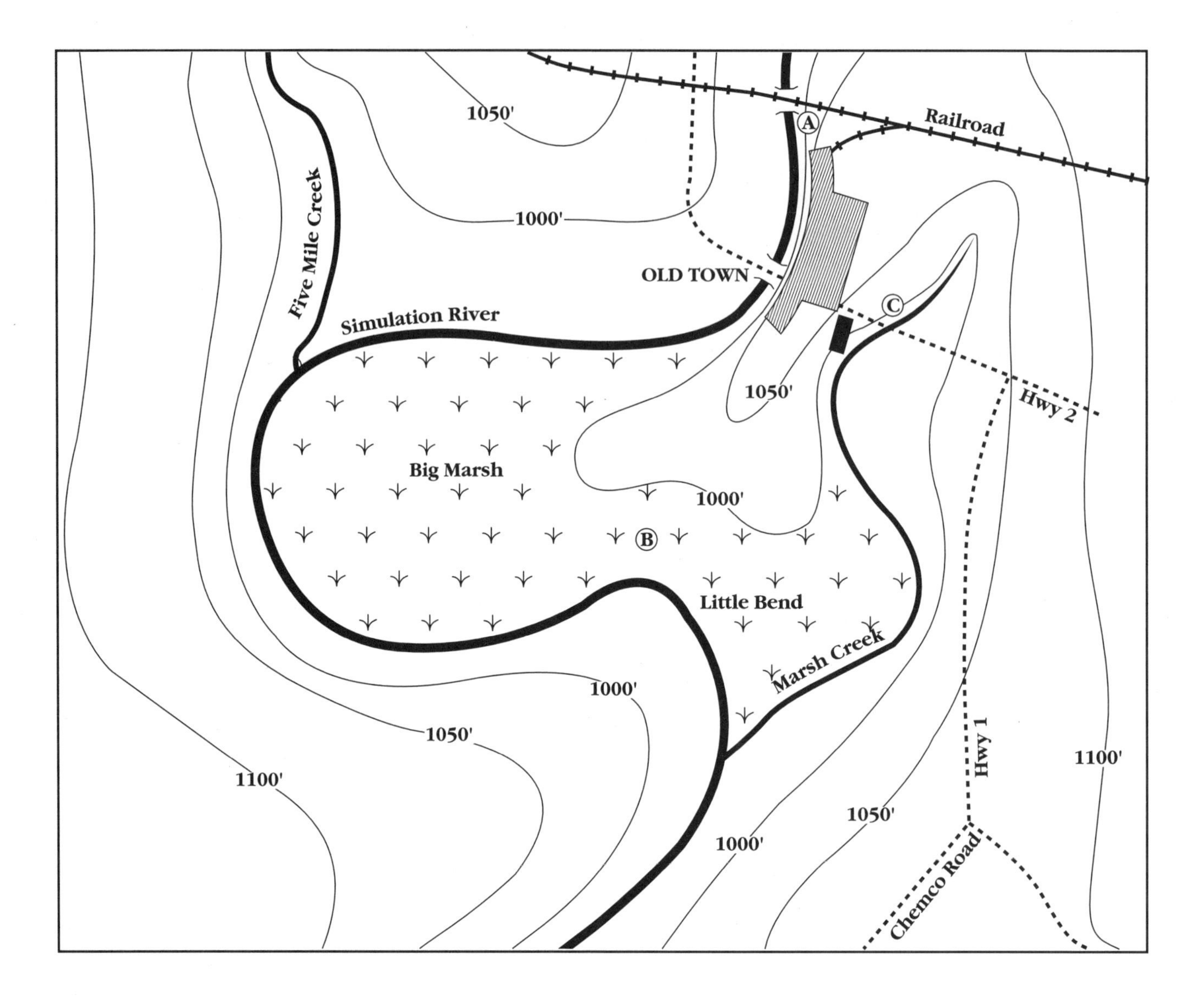

Scale: 1 inch = 1 mile

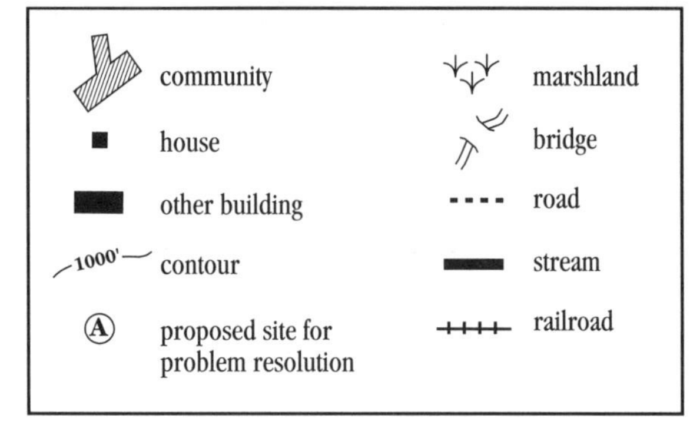

MAP OF REGION C

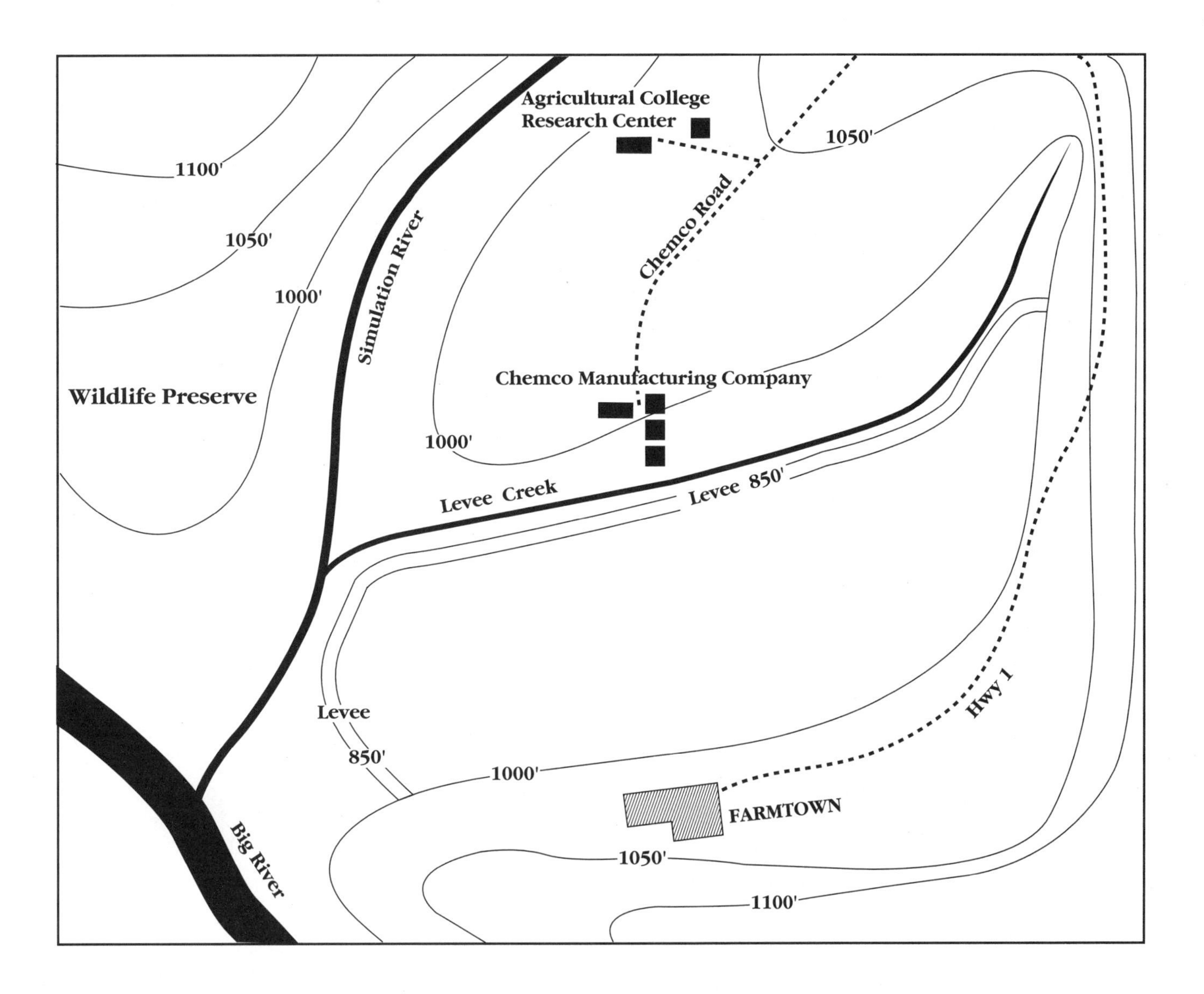

Agricultural College
Research Center
1050'
1100'
1050'
1000'
Simulation River
Chemco Road
Wildlife Preserve
Chemco Manufacturing Company
1000'
Levee Creek
Levee 850'
Levee
850'
1000'
Hwy 1
FARMTOWN
Big River
1050'
1100'

Scale: 1 inch = 1 mile

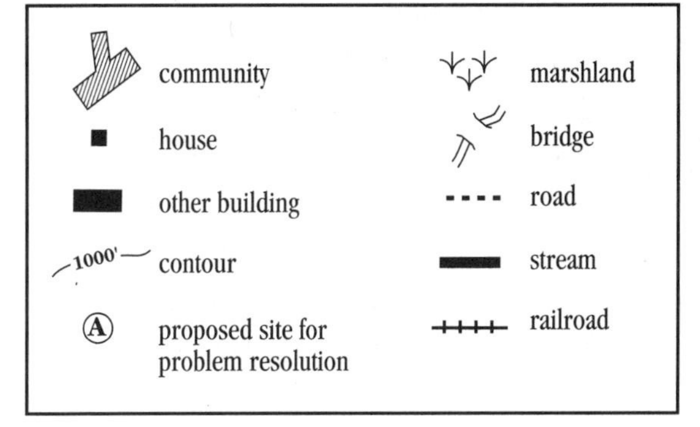

community
house
other building
1000'
contour
proposed site for problem resolution
marshland
bridge
road
stream
railroad

MAP OF REGION D

Name

INFORMATION 5.1

Human Modifications of the River

The physical environment does not compel humans to use it in a predictable fashion. The perception and subsequent use of an environment is influenced by cultural values, technological capabilities, and human choices. Thus, a variety of human activities can be identified in very similar physical environments. The physical environment does pose some limitations and challenges to human uses, but each landscape displays unique human features owing to human choice. The ability to change the environment is limited by available technology and funding.

Human Interactions with Rivers

Human interactions with natural river environments range from simple activities, such as viewing rivers, to complex activities, such as making permanent, large-scale changes to the river. Decisions to modify rivers are based on the belief that the changes will make the rivers more useful for human activities. Natural advantages may be capitalized on to improve economic activities. A dam can harness the energy of fast-flowing water to provide hydroelectricity. Locks and dams can improve low-cost water transportation for bulky commodities. Levees and dams can control flooding to protect valuable agricultural land on floodplains and prevent losses to residences and businesses. Canals may be constructed to link waterways.

Human Modifications of Rivers

The changes made by humans to the earth's natural features are called **environmental modifications.** As populations have grown and technological capabilities have expanded, humans have made many modifications to rivers throughout the world.

In North America, the St. Lawrence River and the Great Lakes have been modified to permit the passage of large ships. Dredging has increased the depth of the navigation channel to 8.1 meters (27 feet). The construction of 16 locks has helped overcome the 174-meter (580-foot) change in elevation. Eight of these locks allow ships to bypass Niagara Falls. Now large ocean-going vessels can be accommodated, and cities on the shores of the Great Lakes have prospered because of their waterway link with the ocean and the world beyond.

The gradient of the Upper Mississippi River between St. Louis and Minneapolis–St. Paul has been modified to facilitate the passage of barges. The construction of 27 locks and dams has reduced the Mississippi to a series of pools, allowing barge tows to move millions of metric tons of grain, coal, ore, and other heavy goods up and down the river. Dredging and wing dams maintain the navigation channel. Canals link the Mississippi, Ohio, Tennessee, Arkansas, and other rivers to the Great Lakes.

Russia's Volga River once emptied naturally into the Caspian Sea, an internal drainage system. The construction of the Volga–Don Canal linked the Volga with the Black Sea, providing an outlet to the Mediterranean Sea and oceans beyond. To the northwest, the Volga is connected with the Baltic Sea by the Volga–Baltic waterway. Dams with locks facilitate shipping and provide hydroelectricity to cities along the Volga. Other canals crisscross Europe, particularly on the relatively flat terrain of the North European Plain.

In China, the Grand Canal—extending more than 1600 kilometers (1000 miles)—connects the country's largest rivers. The Gezhouba Dam on the Chang Jiang (Yangtze River) controls flooding and produces electricity. In the same region of narrow valleys and gorges, the Three Gorges hydroelectric project is underway just downstream from the heavily populated Sichuan Basin. When completed, the Three Gorges dam will be one of the largest in the world: 180 meters (600 feet) high and 2.4 kilometers (1.5 miles) long. The project will require the resettlement of more than a million people to make way for the reservoir that will stretch 600 kilometers (372 miles) upstream. Three Gorges far exceeds other major hydroelectric projects, such as the Itaipu (Brazil–Paraguay) and Yacyreta (Argentina) dams on the Paraná River and the planned Turukhansk dam on the lower Tunguska, a tributary of Russia's Yenisey River.

In dry regions, dams are built to conserve water for domestic and agricultural uses. The Colorado River provides water for the southwestern United States, some water being piped as far away as Los Angeles. In Australia, tunnels bored through the Australian Alps divert the flow of the Snowy River into the Murray–Darling basin. In Central Asia, water from the Amu Dar'ya is diverted for irrigation.

Floodplains are what the name suggests. In the natural order, they serve as temporary reservoirs for river flood water. However, a floodplain's alluvial soils have proven irresistible to humans who want to use them for agriculture and, increasingly, for the construction of factories, highways, and residences. Where humans have chosen to live on floodplains, they have constructed levees to protect themselves and their properties during times of flood. Sometimes, as in the lower Mississippi and in China's Huang valley, repeated raising of levees has elevated the river channel above the floodplain.

Construction and maintenance of levees are costly operations. Furthermore, the feeling of security offered by a levee may be deceiving. In 1993, residents

DRIFTWOOD

Most of the world's largest rivers have dams. Outside of the Arctic, only two major rivers do not have dams: the Zaire in Africa and the main stem of the Amazon in South America.

DRIFTWOOD

"Freshwater systems tend to be the first habitats to experience degradation because humans congregate along waterways."

From *World Resources 1994–95*

of the Upper Mississippi and Missouri floodplains watched in horror as floods crested or broke levees and washed away their homes.

In the aftermath of the floods of 1993, human management of rivers has come into greater question. Should levees be rebuilt? Are levees really effective against the floods, or do they contribute to flooding by restricting river channels? Should more dams be built on tributaries? Should wetlands be restored? Should building be permitted in the floodplain?

Negative Effects of Human Modifications of Rivers

The modifications of rivers are done with good intent; they are designed to improve the lives of the people who live nearby. Unfortunately, the modifications may also produce negative effects. Certainly they alter the natural order. They may also destroy natural habitats and migration routes, as in the case of the decimated salmon populations in the Columbia and other rivers. So much of the Colorado River's water is diverted for the growing population and farming in naturally arid California and Arizona that only a trickle reaches the Gulf of California. A similar condition exists in the Murray–Darling system in Australia. Because of the diversion of water from the Amu Dar'ya in Asia, the Aral Sea is an ecological disaster. Fishing boats sit marooned on a dry sea bed that was once covered with water—reminders of the time when the Aral Sea was twice its current size and fishing was a thriving enterprise.

Denuded hillsides, paved surfaces, and rooftops increase runoff into streams and rivers. Poor agricultural practices accelerate erosion and increase sedimentation. The discharge of municipal, industrial, and other wastes into rivers pollute the water and destroy aquatic life.

Modifications made to improve human life may adversely affect the environment. Cutting down trees provides building materials and fuel but accelerates runoff and erosion. Harvesting of crops provides food but leaves barren fields open to erosion. Paving highways improves transportation, constructing buildings provides shelter, and providing paved parking lots improves business, but the impervious surfaces of these human features increase runoff. Discharging waste into streams is cheaper than other disposal methods, but it may harm or destroy aquatic life, destroy habitats, and make the water unusable downstream.

Not even the most ardent environmentalist expects all rivers to be returned to their natural state. On the other hand, past human efforts to control the environment should provide lessons for future planning. Bowing to the narrow viewpoints of just a few groups is not workable. Coexistence with the river requires a comprehensive view.

Driftwood

"The great red hills stand desolate, and the earth has torn away like flesh. The lightning flashes over them, the clouds pour down upon them, the dead streams come to life, full of red blood of the earth. Down in the valleys women scratch the soil that is left, and the maize hardly reaches the height of a man. They are valleys of old men and old women, of mothers and children. The men are away, the young men and the girls are away. The soil cannot keep them any more."

From *Cry, The Beloved Country* by Alan Paton

Questions

Write your answers on a separate piece of paper. Use complete sentences.

1. What human factors influence modification of the physical environment?
2. What challenges do rivers pose for human use?
3. How do humans modify rivers to serve human needs?
4. How can well-intentioned changes to rivers produce negative results?

Driftwood

"To waste, to destroy, our natural resources, to skin and exhaust the land instead of using it so as to increase its usefulness, will result in undermining in the days of our children the very prosperity which we ought by right to hand down to them amplified and developed."

From Message to Congress by Theodore Roosevelt

Name ______________________

Making Decisions About the Environment

People's opinions vary about the best uses and non-uses of rivers and other features of the physical landscape. Given the great variety of possibilities and human needs and wants, this should not be surprising. Arriving at a plan for the best relationships between humans and the physical environment requires that all points of view be considered.

Special-interest groups are people bound together to express their collective preference for particular interests. These preferences and interests may range from enormous environmental modifications on one hand to **conservation** measures that preserve or restore the natural environment on the other hand. Some groups may favor a project in general but oppose its development in the vicinity of their homes. This attitude is referred to as **NIMBY,** an acronym for Not In My BackYard.

> **DRIFTWOOD**
>
> **"The future of mankind lies waiting for those who will come to understand their lives and take up their responsibilities to all living things."**
>
> **From *God is Red* by Vine Victor Deloria, Jr.**

Identifying Problems and Solutions

An example of a major problem that affects the environment is the creation and disposal of **waste,** the discarded materials that often contribute to pollution. The average United States citizen produces more than 1.8 kilograms (4 pounds) of waste per day or about 660 kilograms (1450 pounds) per year, posing enormous problems for disposal. Many people advocate reducing the amount of waste by **recycling,** or collecting and reusing waste materials. Some communities provide separate trash containers for recyclable items, such as plastics, newspaper, glass, and metals.

Most solid waste is covered with earth in specially prepared sites called **sanitary landfills,** which are designed so that harmful materials will not leach out and contaminate soil, groundwater, and streams. Landfill sites are in short supply due to increasing amounts of solid waste, a shortage of suitable sites, and citizens' opposition to the construction of new landfills near their homes and communities.

Another method of disposal involves burning wastes in specially constructed incinerators. This method is expensive because it requires that care be taken not to release pollutants into the air. Also, many citizens do not want incinerators located near their homes.

Some waste continues to be dumped into rivers, streams, and other waterways, which carry it away from local communities, but this approach may pose health hazards related to water consumption, swimming, and eating

> **DRIFTWOOD**
>
> **"Indeed one of the most alarming aspects of the chemical pollution of water is the fact that here in river or lake or reservoir, or for that matter in the glass of water served at your dinner table, are mingled chemicals that no responsible chemist would think of combining in his laboratory."**
>
> **From *Silent Spring* by Rachel Carson**

aquatic life exposed to the wastes. Water quality in many U.S. streams has improved in recent years, perhaps owing to greater public awareness of problems associated with dumping wastes into rivers, streams, and other waterways.

Recognizing Conflicting Interests

Economic and regional interests may influence decisions about the environment. On some rivers, constructions of dams with hydroelectric generators can economically produce electricity to provide modern conveniences and run our society. Barge traffic can carry tremendous loads of resources, such as lumber to build homes; petroleum to make gasoline, plastics, and fertilizers; iron ore to make steel; and coal to generate electricity. Accommodating such barge traffic, however, may require installing locks and dams on the river and deepening the river channel. Wetlands restoration, though possible, may cause the loss of productive farmland. Building or rebuilding houses on floodplains may lead to construction of protective levees.

One **region** (an area differentiated from other areas according to specified criteria) may have very different interests and preferences from other nearby regions. For example, people in one region along a river may derive their livelihood from the timber industry. People in an adjacent region may depend on the aquatic life and natural beauty of the river to attract tourists and nature enthusiasts. No doubt these two groups will view the river very differently and will have conflicting points of view about how the river should be used or conserved.

Making Environmental Decisions

Environmental decisions require input from all points of view. A final decision is seldom made by simply selecting a good choice over a bad one. Furthermore, a decision made by a "show of hands" may not reflect the best sustainable solution that promotes long-term coexistence with the river. A popular position about the "right" course of action may serve only immediate needs.

If a proposed environmental modification is likely to have a major impact, those responsible for decisionmaking should carry out an **environmental impact study** assessing the changes to the physical environment (and human landscape) that those human modifications may cause. The purpose of an environmental impact study is to provide an objective evaluation of the effects of a change. An environmental impact statement (EIS) must not only provide thorough information regarding the environmental impacts of the proposed project but also identify and explore all reasonable alternatives to the proposed project. An EIS must analyze scientifically the direct and indirect

Driftwood

"We abuse land because we regard it as a commodity belonging to us. When we see land as a community to which we belong, we may begin to use it with love and respect."

From *A Sand County Almanac* by Aldo Leopold

Driftwood

The creation of a shipping lane along the entire length of the Paraguay–Paraná river system in South America threatens the Pantanal, one of the largest wetlands in the world. In the rainy season, the Pantanal absorbs vast quantities of water, preventing flooding of the lower reaches of the river system where 25 million people live. The Pantanal is also home to hundreds of species, including endangered mammals, such as the marsh deer, the giant otter, the jaguar, and the giant anteater.

consequences of the proposed action. The EIS should include reporting on field studies conducted to evaluate potential changes to ecosystems, as well as regulations that may affect the project, such as changes in water quality and the impact of such changes.

An EIS must also consider the effects of the proposed project on the human landscape. For example, would the change involve relocation of residences? Would it affect agriculture and mining land uses? Would it require relocation of electric lines, pipelines, water lines, or telephone lines? Would it involve new roads? The specific questions vary to match the situation, but, overall, an environmental impact statement should provide enough data and analysis that people can weigh the pros and cons of the proposed change and make an informed decision.

Questions

Write your answers on a separate piece of paper. Use complete sentences.

1. In your own words, write a brief definition of each of the following terms:
   a. special-interest group
   b. conservation
   c. NIMBY
   d. waste
   e. recycling
   f. sanitary landfill
   g. region
   h. environmental impact study
2. List at least two ways humans have modified rivers, and record advantages and disadvantages of each example.
3. Why do people disagree about how rivers and other natural systems should be treated? Use examples to explain.
4. List several pairs of types of opponents whose opinions might differ on how humans should treat the river.

Driftwood

"A land ethic for tomorrow should be as honest as Thoreau's *Walden*, and as comprehensive as the sensitive science of ecology. It should stress the oneness of our resources and the live-and-let-live logic of the great chain of life. If, in our haste to 'progress,' the economics of ecology are disregarded by citizens and policy makers alike, the result will be an ugly America."

From *The Quiet Crisis* by Stewart Lee Udall

Name

Simulation of Environmental Decisionmaking

Purpose

To role-play making decisions about land and water use.

Background

In this activity, you will engage in a **simulation,** a situation created to imitate reality so you can explore more directly the interactions of people and natural processes. You will be assigned to one of four groups; each group represents a different community faced with a water- and land-use problem. Each member of your class will assume an assigned role within one of these hypothetical communities. Through **role playing**—assuming the identify and characteristics of hypothetical individuals in those communities—group members will study the problem faced by their community and arrive at a decision.

The activity has three parts. Read and complete one part before reading the next part. In Part A, you will become acquainted with your hypothetical region, the problem faced by the region, and the roles that you and others will play. In Part B, your group will simulate a community meeting in which members present different points of view and the group makes a decision about how to resolve the region's problem. In Part C, you will see what decisions the other three regions made, and you will prepare an environmental impact statement.

Materials

Per group

- large envelope containing:
  - laminated regional description
  - laminated regional map
  - photocopy of regional map per group member
- laminated role cards rubber-banded together (will be distributed later)
- laminated regional problem (will be distributed later)

For the teacher

- 1 photocopy of each regional map
- masking tape

PART A. Simulation of Community Decisionmaking

Procedure

1. Join a specified group as instructed by your teacher.
2. When your teacher distributes to your group a map and a description of your hypothetical region, examine these materials and answer the first question in the Observations section.
3. When your teacher distributes the cards that describe the roles that each member of your group will play, choose roles as directed by your teacher.
4. Playing your role, work with your group to complete the observations, analyses, and conclusions that follow, written or orally, as directed by your teacher.

Observations

1. Locate the features of your region on the map. Briefly describe your community and region.
2. Read your role and write a description of it.
3. Have each person in your group read aloud his or her role. In your own notes, record descriptions of each role.

4. Read the problem your community faces, and write it in your own words.

Analyses and Conclusions

1. Based on your role, write down two solutions to the problem that you would consider acceptable and one that you would consider unacceptable.
2. Briefly note three arguments that could be used to win others over to your point of view.
3. As a group, hold a "community meeting" to discuss the solutions and supporting arguments suggested by your group. As others voice their opinions, place an "X" on your map for each objection to the use of that site to solve the problem. As a group, decide on a solution that everyone in your group finds acceptable. Write the solution your group finally agreed on.
4. How did your group reach a resolution to the problem? What arguments were most convincing?

PART B. Simulation of Regional Decisionmaking

Procedure

1. Select a member of your group to briefly describe your region and community and the roles played, and to present your community's problems addressed, choices considered, and the group's solution.
2. As you listen to the presentations, record each group's regional description, problems, and solutions.
3. Participate in a regional meeting in order to complete the observations, analyses, and conclusions that follow.

Analyses and Conclusions

1. Considering the overall impact of these solutions, what was the net result of the four decisions?
2. How does the impact of the combined decision differ from the decisions made by the individual regions? How can problems solved in one region affect other regions?
3. How can such problems be resolved to the satisfaction of all?
4. Participate in a "regional meeting" in which the interests of each group and the conflicts between different groups are discussed. As you hold this discussion, refer to a map that shows all the regions. Try to arrive at a solution that includes all four regions.

PART C. Environmental Impact Study

Procedure

1. Listen to the proposed change described by your teacher. Take notes as needed.
2. Respond to the analyses and conclusions that follow.
3. Using those answers, work with your group to prepare an environmental impact report that predicts the expected effects of the proposed human alteration to the natural environment.

Analyses and Conclusions

1. List as many changes as you can that the proposed construction is likely to cause in the physical landscape. Include environmental changes and human modifications that are likely to occur as people adapt to their new environment.
2. List as many changes as you can that the proposed construction is likely to cause in the human environment.
3. How will humans adapt to the new environment?

Creating an Environmental Impact Report

1. Based on your observations, list the advantages of the proposed construction.
2. List the disadvantages of the proposed construction. Remember, a particular change may be considered beneficial from one point of view and nonbeneficial from another. Try to view possible changes from a variety of perspectives.
3. Together with other members of your group, prepare a written environmental impact report that lists and explains the major impacts the project will have on the physical and human environments of the

watershed. Though you cannot anticipate all effects, you can predict many with accuracy.

4. Select a member of your group to present your report to the class. Be prepared to assist by offering explanations in the discussion that will follow.

Performance Criteria

The successful environmental impact report:

- Identifies changes to the physical elements of the environment.
- Identifies changes to the human elements of the environment.
- Proposes alternatives to the proposed project.
- Projects potential adaptations of humans to the changed environment.
- Presents information in a well organized manner.

Keeping Your Journal

1. Consider the problems of your local community. Do the problems involve the river, wetlands, or land use? What issues are involved and what are the conflicting points of view?
2. How has your local river been modified? Are the changes beneficial or harmful?
3. How are decisions made in your community?
4. How do people in the local community interact with the river? Select an activity and describe it in detail.

TEACHER NOTES

LESSON 6

What Can Be Learned From a Field Study?

Focus: Field Study

In this lesson, students will apply the concepts they have learned in class to an outdoor field study of a river or stream environment. First, they will compare their direct observations of physical and human features with symbolic representations on a topographic map of the site. Then they will demonstrate a knowledge of spatial concepts by sketching and labeling a map of the study site. Finally, they will express what they see, discuss their observations, and write about the field site in their journals.

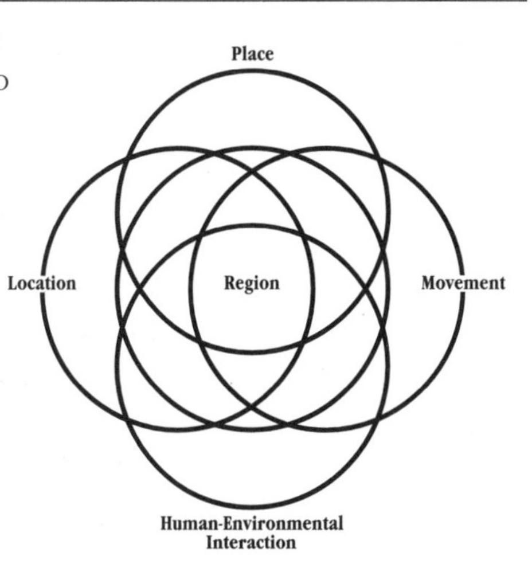

Perspectives and Geography Standards

Perspectives: Spatial; others, depending on study site

Geography Standards Receiving Primary Emphasis

This lesson primarily promotes mastery of the following standards:

3. How to analyze the spatial organization of people, places, and environments on Earth's surface
4. The physical and human characteristics of places
14. How human actions modify the physical environment

Geography Standards Receiving Secondary Emphasis

Several other standards receive secondary emphasis during specific activities: students sketch maps of the observation site (Standard 2), identify regions and ecosystems and research evidence of change (Standards 5 and 8), and evaluate potential land and water use conflicts (Standards 13 and 18).

Learner Outcomes

Students will:

1. Make field observations of natural and human features and record these features on a sketched map of the site.
2. Compare primary and secondary sources of information.
3. Analyze the interrelationships among physical and human features.

Time

Three class periods of 40–50 minutes per period, plus a half-day site visit for field observations

DAY 1 (SEVERAL DAYS BEFORE REST OF LESSON):

Student Information 6.1 Safety Guidelines and Contract (partial class period, several days before the field trip; also distribute permission forms)

DAY BEFORE STARTING LESSON:

Assign as homework Student Information 6.2: Field Study in the Geographic Process

DAY 2: Discuss Student Information 6.2: Field Study in the Geographic Process
Student Activity 6.3: River or Stream Field Study, Part A (and corresponding Teacher Guide)

DAY 3: Student Activity 6.3: River or Stream Field Study, Part B (half day)

DAY 4: Complete and review Student Activity 6.3: River or Stream Field Study, Part C

Advance Preparation

Read Lessons 6 and 7. Prepare to distribute to each student Student Information 6.1 and 6.2 and Student Activity 6.3. Gather all necessary materials for this lesson. As an alternative to obtaining and laminating one copy of the study site topographic map per two students, you may photocopy and distribute just the portion of the topographic map representing the study site area.

Prepare for field study, as detailed in Teacher Information 6.1: Field-Study Guidelines. If desired, this may include securing the assistance of a conservation officer or some other field expert to speak to students and answer questions at the field site.

Select and duplicate for distribution to students brief descriptions of rivers and river-related activities that illustrate contrasting writing styles. For sources, see *Rivers Language Arts,* other textbooks, *National Geographic,* or the writings of such authors as Edward Abbey, Mark Twain, and Henry David Thoreau.

Safety and Waste Disposal

Review Student Information 6.1: Safety Guidelines and Contract with your students before the field trip and again on the day of the field trip. Emphasize that no wastes should be left at the study site. The number of life jackets, tow lines, and wading boots provided should reflect the character of your local site and the number of students who will have direct contact with the water. (If students will be performing water-quality tests, make sure students know proper procedures for handling and disposal of test materials. For further details, see Safety Guidelines and Contract in *Rivers Chemistry.*)

Materials

Student Activity 6.3: River or Stream Field Study

Per student

photocopies of writings about rivers and river-related activities
Student Information 6.1: Safety Guidelines and Contract, signed
Student Information 6.2: Preparation for Field Study
permission form (signed by a parent or guardian)
ruler
journal
notebook
clipboard or other firm writing surface
pencil
several sheets of lined and unlined paper
camera and film (optional)
appropriate clothing (see Student Information 6.1)
insect repellent, sunscreen, medication, sunglasses as appropriate
bottle of drinking water

Per two students

laminated topographic map of the field site (7.5-minute series; 1:24,000 scale is preferable) (or one photocopy per student of field-site portion of topographic map)

Per group

Celsius thermometer, alcohol-filled with metal jacket, or electronic
watch

Per class

other maps and materials, such as wetlands maps, plat books, EPA reports, conservation department information
compass
first-aid kits
life jackets and tow lines for deeper-water sites
wading boots for shallower-water sites
protective gloves
bottled water for washing
soap and towels
paper cups
large plastic bags for collecting waste

Vocabulary

analysis
personality of place
synthesis

Background for the Teacher

Field studies are an important part of developing geographic understanding. They offer an opportunity to observe geographic elements firsthand. Unlike reading a textbook or studying a map, in which selected topics have been generalized for easier understanding, in the field students confront an environment without labels. To make sense of what they observe, students must apply concepts learned in the classroom to real-world situations.

Moreover, investigators, whether students or professionals, may not be able to find sufficient information about a particular local area to study it in detail; firsthand observation remains the only method by which to gain such understanding.

Finally, the "feel" an investigator obtains from direct observation, although unscientific, may make essential contributions to understanding a place. Purposeful (as opposed to casual) observation may stimulate curiosity and reveal relationships that students could not imagine by simply reading about the river in a textbook or hearing another person's description.

As an optional activity, students may investigate water quality at the field site. When students at a field site look for indicators of human effects on the water, they often want to look at the water itself for clues. Scientists determine the quality of the water in a river or stream by using a series of nine specific water-quality tests. The results of most of these tests are strong indicators of human-environmental interactions. Running and analyzing these tests gives students a powerful connection to the waterway. You may want to work as a team with a chemistry teacher at your school who is using *Rivers Chemistry*. For some sites, other students have already contributed water-quality test data to the Rivers Project database. Information on accessing the database is on page x.

Even students (and teachers) who have not had training in chemistry can perform some of these tests in conjunction with a field-site visit. Complete background information, materials lists, and student activities on water-quality testing are provided in *Rivers Chemistry*. If students are to conduct water-quality monitoring tests, allow additional time in the classroom to discuss safety and to practice the tests, at the field site to conduct the tests, and in the classroom after the tests to analyze the results. For tests involving test kits, necessary safety precautions and materials can also be found in the kit manuals (see Teacher Notes, Extending the Lesson.)

Introducing the Lesson

1. At least one week before the field trip, discuss Student Information 6.1: Safety Guidelines and Contract with students. Begin with a brief overview of Lesson 6. Review with students your school's procedures for field trips. Have each student take home Student Information 6.1 contract and a field-trip permission form for signature by a parent or guardian. Make sure all students have returned signed permission forms and safety contracts before the trip.
2. The day before beginning Lesson 6, assign as homework Student Information 6.2: Field Study in the Geographic Process.

3. Orally review Student Information 6.2. (Sample answers are in Appendix B.) Be sure students understand that in the field study they will be applying concepts they have learned in other lessons and gathering information for Lesson 7 activities.
4. Describe the study site to be visited. Review topographic and other maps that may facilitate student recognition of features in the field.
5. Distribute Student Activity 6.3: River or Stream Field Study. Preview it and have students complete Part A. (Use corresponding Teacher Guide.) Distribute or present information on any additional special tasks students will do during the field study. Remind students to bring Student Activity 6.3 on the field site visit.
6. Assign as reading homework the photocopied descriptions of rivers and river activities as preparation for the journal entry that students are to write at the field site.
7. Briefly explain the brochure activity of Lesson 7. Instruct students that they should be thinking of topics of interest to them and gather information during the field study for use in their brochures. Suggest that they may want to take photographs or make sketches during the field study for use in the brochure project.

Developing the Lesson

1. On the way to the field site, review once more all the safety guidelines with students.
2. At the field site, have students complete Part B of Student Activity 6.3. (Use corresponding Teacher Guide.)

Concluding the Lesson

1. In the classroom (or as homework), have students complete Part C of Student Activity 6.3.
2. Use students responses to Student Activity 6.3 as a basis for discussion of the field study. Emphasize the sharing of information gathered at the field site. Ask students how their impressions changed as a result of their first-hand observations of the river and its environment. The review may be conducted by the class as a whole or by small groups followed by a general class discussion.
3. When discussing student journal entries, have students compare and contrast their entries about the river with the writings you distributed to them prior to the field trip.

Assessing the Lesson

1. Collect the observations and writings done as part of Student Activity 6.3 for assessment.
2. The brochure that students will prepare in Lesson 7 will be partly based on students' field observations, so the brochure project also can be used to assess this lesson.

Extending the Lesson

1. If appropriate to your field site, conduct a walk along the river to observe a variety of physical phenomena such as landforms, erosion, vegetation, and habitats, and to see how humans have changed the environment. For suggestions on conducting a streamwalk, read *Streamwalk—A stream monitoring tool for citizens.* United States Environmental Protection Agency, Water Division Region 10, 1200 Sixth Avenue, Seattle, WA 98101.
2. Plan a river or stream cleanup activity as part of your field-site visit. Contact your local Department of Transportation or Department of Conservation.
3. Return to the river or stream for a follow-up study. Assessment of the river study may provoke additional topics that the class may wish to pursue.
4. Involve students in water-quality monitoring. For information on monitoring water quality, consult *Rivers Chemistry* and the following publications:

   McDonald, Brook, William Borden, and Joyce Lathrop. *Citizen Stream Monitoring—A Manual for Illinois.* Illinois Department of Energy and Natural Resources, August 1990.

   Mitchell, Mark K. and William B. Stapp. *Field Manual for Water Quality Monitoring,* 9th ed. Dexter, MI: Thomson-Shore, 1995. Order from: William B. Stapp, 2050 Delaware Ave., Ann Arbor, MI 48103.

TEACHER

INFORMATION

6.1

Field-Study Guidelines

Field studies, particularly to sites around a body of water, require careful planning and management. Here are steps to make *Rivers Geography* field-site visits safe and successful. Adapt this list to fit your own circumstances and preferences.

Before the Field Study

__ Secure school administration permission for field-site visit.
__ Contact a conservation officer or another expert to help with site selection and to talk to students and answer questions at the field site.
__ Select a study site and alternative study site.
__ Secure permission from landowner; if public land, contact the administrator to alert him/her to your plans.
__ Inspect the site personally.
__ Publicize the event with local newspaper, television, and radio.
__ Invite the school yearbook or newspaper photographer to accompany the group.
__ Arrange for field-site visit transportation.
__ Arrange for appropriate adult supervision on trip.
__ Invite parents and administrators to come along on the trip.
__ Prepare directions for drivers.
__ Develop an activity plan and timetable for the field-site day.
__ Arrange for lunches or snacks, drinking water, and rest rooms.
__ Obtain the following items to take on the trip:
  __ first-aid kits
  __ insect repellent
  __ life jackets and tow lines (for deeper-water sites)
  __ rubber wading boots (for shallower-water sites)
  __ emergency telephone numbers
  __ protective gloves
  __ sunscreen
  __ safety ropes
  __ bottles of water for washing
  __ soap and towels
  __ water for drinking
  __ paper cups
  __ plastic bags for collecting waste

__ Distribute field-site visit permission forms to students, and receive signed forms several days before the field-site visit. (Permission forms should include request for information about student allergies and other medical needs.)
__ Distribute Student Information 6.1: Safety Guidelines and Contract.
__ Give instruction on field-site clothing and personal items (not only those listed in Student Activity 6.3 but also lunch, notebook, and so forth as appropriate).

Day Before the Field Trip

__ Make sure all signed student permissions and safety contracts have been turned in.
__ Review Student Information 6.1: Safety Guidelines and Contract with students.
__ Review activities that will take place during field-site visit, including extra activities, such as writing and litter pickup.
__ Assign any specific tasks that are to be done by students.
__ Caution those students who plan to take binoculars, cameras, tape recorders, or video cameras to carry them in protective cases.
__ Each group assembles and prepares materials and equipment for their specific test or tests to be done at the river.
__ Pack all field-study and safety materials and equipment.
__ Assign students to complete reading of Student Activity 6.3.

Field-Trip Day

__ Distribute directions to drivers.

__ Assign students to load equipment and materials for transportation to field site.

__ Assign responsibilities to supervising adults.

__ Review Student Information 6.1: Safety Guidelines and Contract with students again during drive to field site.

__ Distribute safety and field equipment and materials.

__ Students do assigned observations and activities.

__ If students have completed assigned activities, have them help other students, do waterside cleanup, or write in their journals.

After the Field Trip

__ Students complete observation reports and journal entries for Student Activity 6.3.

__ Clean, store, and return equipment.

__ Dispose of waste.

__ Send written articles about the field study to school and local newspapers.

__ Send thank-you notes to those who assisted with the project.

__ Review procedures and revise as necessary for the next field study.

T E A C H E R

GUIDE

River or Stream Field Study

For Student Activity
6.3

Purpose

In parallel with Student Activity 6.3: River or Stream Field Study, to provide specific teaching steps and sample answers that facilitate students identifying in a real environment the geographic concepts learned in the classroom.

Background

During this activity, students will study a particular river or stream field site on a topographic map, prepare for a site visit, make observations at the field site, and, finally, analyze and draw conclusions based on that information.

This Teacher Guide concentrates on the activities and teaching steps that apply specifically to guiding student learning during the field study. Specific planning and management activities involved in teaching an activity involving a field-site visit are included in the preceding Teacher Information 6.1. For the materials list for this activity, see Student Activity 6.3 or the Teacher Notes for Lesson 6.

Safety

Follow all field safety procedures presented in Student Information 6.1: Safety Guidelines and Contract.

Procedure

PART A. Before the Field Study

1. Review the Safety Guidelines. Sign the Safety Contract and give it to your teacher. **Collect signed safety contracts.**
2. Have a parent or guardian sign your permission form; give the signed form to your teacher. **Collect a signed permission form from each student.**
3. Read the entire activity.
4. Carefully examine a topographic map of the study site and assess the features and relationships you expect to observe at the site.
5. On a separate piece of lined paper, create the labels and entries listed in the Observations section of this sheet.
6. Determine the absolute location of your field site to the nearest degree; record it in your own Observations section.
7. Determine and record the location of the field site relative to your school, using cardinal or intermediate compass directions, and record in your Observations section.
8. Calculate and record the distance (straight line and by road) of the study site from your school. **Give students class time to complete steps 3 through 8.**

9. On the day of the field study, double-check the materials list and safety guidelines. **Complete all field-site visit preparations listed in Teacher Information 6.1.**

PART B. During the Field Study

Monitor students carefully to make sure they are following all appropriate safety procedures. Student groups should be supervised by a responsible adult at all times. Students should complete Part B at the field site, including the Observations section and the field-site sketches. Suggest that students look around to get an overall impression before writing their general description of the site.

10. At the field site, complete the Observations section for this activity in your *Rivers Geography* notebook. Be sure you record sufficient information to answer the Critical Thinking Questions. Your answers to the questions will form the basis for classroom discussion following the field study.
11. On a piece of unlined paper, sketch in pencil the river or stream at the study site. Label the natural and human features of the area. Include any of the following as appropriate:
    a. Natural features: estimated width of river (or stream) and floodplain, islands, rapids, sand or gravel bars, oxbows, wetlands, tributaries, springs, obstructions in river or stream channel, vegetation cover.
    b. Human features: houses and other buildings, roads, railroads, bridges, power lines, dams, boat ramps, riprap.
12. Compare your topographic map with the landscape you see at the study site. Record on your sketch map or on your field trip observations any omissions, corrections, or additions.
13. Add to your record any other observations you feel are worth noting, particularly those relating to the possible topics you have selected for possible further development in Lesson 7. Keep in mind the themes of geography.
14. If you wish, make sketches or take photographs to support the development of your possible topic or topics of particular interest.
15. In your journal, write your impressions of the river or stream. What is its "personality?" What do you feel like when you are at the site? Record all your impressions, including your feelings about the river or stream. Remember that you will use this information in Lesson 7.
16. As necessary, read the remaining questions on this sheet and take pertinent notes at the field site.

PART C. Synthesizing the Field Study

Tell students whether they should do Part C of Student Activity 6.3 at the field site, in the classroom, or as homework. When students have completed this work, discuss it.

17. After the field-site visit, before the next class session, complete on a separate piece of paper the sections labeled Analyses and Conclusions and Critical Thinking Questions.
18. Complete the Keeping Your Journal segment of this activity.

Observations

1. Site Identification
   River or Stream Name: ________________
   Location: Absolute (to nearest whole minute) ________
   Relative to school (using cardinal or intermediate compass directions)
   ____
   Distance from school: straight line: _______ km (_____ miles)
   by road: _______ km (_____ miles)
2. Conditions at Site
   Temperature: ___________ Date: _______________ Time: __________
   Cloud cover: ____________________ Weather: ____________________
3. General description of the site:
   Description will vary with site and time of year. It should contain references to many of the natural and human features studied earlier.
4. Observable human uses and activities or evidence of such (List types and numbers): *Students should easily identify many human features. Identification of some uses and former uses may prove more challenging.* **Suggest that students try to envision what the landscape would look like if no humans had ever visited the site.**
5. Observable examples of pollution in the river or stream and in its vicinity: *Amount and type of pollution depend on the site. Obvious forms include litter discarded on the banks or in the water.* **Ask students to consider whether less obvious forms of pollution may be present, such as sedimentation from eroded land, herbicides and pesticides, and industrial and municipal waste discharge. Both the USGS and state EPA monitor such pollutants in streams. Federal USGS and state EPA reports may be available in the school or public libraries and can be used for follow-up studies.**
6. Other observations worth noting, particularly those relating to topics for possible further development in Lesson 7: *Observations will vary with student interests. Specific information may be recorded, such as a detailed analysis of the forest (names of tree varieties, condition of woodlands, descriptions of individual trees), firsthand sensory observations (smells and sounds), or visual impact (appreciation of the natural beauty of the site or revulsion at the presence of trash).* **Some students may become emotionally involved during their observations. You may sympathize, but remind students to maintain an objective viewpoint for their observations; they may write their opinions in their journals.**

Analyses and Conclusions

1. What information about the river or stream site can you determine from the field study but not from the topographic map? *All the senses receive stimulation during a field study, whereas only the visual sense is used in viewing a two-dimensional map. Real phenomena—rather than symbols—are experienced on a field study. Maps simplify the actual landscape and show Earth at a scale much smaller than reality (1:24,000). On a field study, you experience Earth with all its complexity at a 1:1 scale. Depending on the age of the topographic map, students will likely identify new human features. They will also observe phenomena too small or insignificant to appear on the map.*
2. What possible conflicts exist among various users of the river or stream? *Conflicts and examples will vary widely. Generally, students will focus on people who use the river and the river area for recreation, residences, primary economic activities (agriculture, forestry, mining), manufacturing, and trade.*

Critical Thinking Questions

1. How do the natural and cultural features of this location relate to features in other locations? *The river is part of a larger watershed. Landform types and atmospheric conditions may extend far beyond what can be seen from the field site. Transportation routes connect with distant locations. Students may also include references to culture areas and the hydrologic cycle. They may also draw on information covered in previous activities, such as river volume as a consequence of recent rainfall or drought; water quality as a result of upstream pollution; or the imprint on the landscape of a particular culture.*

   You may wish to display regional, continental, and world maps of watersheds, climatic regions, vegetation, and other features as illustrations to accompany classroom discussions.
2. What combination of characteristics make this location unique? *Answers will vary. Some sites may contain few human modifications; others may contain so many that few physical features other than the river and sky are visible. Some may contain one feature that dominates the area: a waterfall or rapids, a marsh, a dense forest, a dam, a marina, or a housing development.*
3. How have the river (or stream) and its immediate environment changed in form and in function? *The river may have been harnessed to serve human needs. Some features change both form and function. For example, a dam changes a free-flowing river to a lake (form) and restricts its erosive action, acts to collect sediments, and serves as a water supply and place for water recreation (function). Riprap and other structures may stabilize the banks (form) and in the process eliminate some riverine life forms and prevent the erosive action of running water (function).*
4. What positive results can you attribute to changes in the river or stream and its environment? What negative results? *Answers will vary. Positive*

changes may include levees and bank stabilization structures to protect human land uses and activities, channelization and dams facilitate the transportation of goods, and boat ramps permit water recreation. Negative changes may include alteration of the river ecology, pollution of the water and adjacent land areas, and the acceleration of sedimentation.

Ask students to assume the perspectives of those persons whose decisions changed the river. What benefits did they expect the changes would provide?

5. How has the significance of the location changed over time? In what ways have local or regional changes altered the relative location of the river or stream? *Improved land transportation may make the river more accessible for human uses, and modifications to control the river may make the area more attractive for human activities. Urban growth, increased recreational demands, and efforts to develop flood-control measures may have increased the desire to "develop" the river and its immediate area.*

   Encourage students to imagine how the river was used in the past, when fewer people lived in the region. You may want to present population figures for selected previous years.

6. How do conflicts over the use of the river or stream arise? Do conflicting uses lead to misuse? How can differences be resolved? *Answers will vary but may include that many interests compete to control the river for their own benefits. What one person or group perceives as beneficial may be viewed by others as misuse. For example, conservationists may want to preserve the river in its natural state, while other interests may promote alteration of the river to make it more useful. Opinions about the best use of the river depend upon the ideas a person or group wishes to promote. Cooperation among interests or legal guidelines are necessary, because the river is finite and cannot meet the desires and needs of all people.*

   Ask students to state what uses of the river would be considered beneficial by each of the following interests: recreational, commercial, industrial, agricultural, and conservationist. How would the uses promoted by each group be viewed by the other groups? Encourage students to share their knowledge about legal controls regarding river uses and the governmental agencies that administer them.

7. How has this field study changed your impressions of the river or stream? What are your strongest positive and negative impressions? *Answers will vary but tend to be strongly positive or strongly negative. Reactions may range from an appreciation of the natural beauty of the river to being disgusted by the pollution.*

   As appropriate, have your student complete the Keeping Your Journal segment. Remind students to describe the "personality" of the study site in their journals, because they will use this information in Lesson 7.

Safety Guidelines and Contract

Fieldwork requires specific kinds of safety. To insure everyone's safety, you must follow these rules in the field. Once you have reviewed these materials, signify your agreement to follow these safety guidelines by signing your name to the Safety Contract at the end of this student information sheet. You should review the rules before you carry out the field study.

1. Stay within the area directed by your teacher. Do not cross private property without the teacher's permission.
2. Always stay with your group.
3. Be on the alert for potentially dangerous wildlife, including bees, wasps, poisonous snakes, and rabid animals. Observe all wildlife from a distance. Check for ticks as you make observations and as you leave the site.
4. Never drink from or wash food in a river or stream.
5. Avoid walking on unstable banks. Exercise caution on slippery rocks or wood.
6. Do not attempt to cross streams that are swift or are above knee height. Never enter the stream without a life jacket. Use a tow line if the water is not extremely shallow. Remember, stream beds may be very slippery and may drop off suddenly.
7. Disturb plants as little as possible. Avoid poison ivy and poison oak and other poisonous or irritating plants. If you are unable to positively identify a plant, avoid it (but remember to find out what it is).
8. If you feel uncomfortable about the stream conditions or surroundings, stop your activity immediately. Safety is more important than collecting information.
9. Wear long pants, preferably jeans. If the day is hot and sunny, wear a hat or cap with a brim. If it is cooler than 70° F or if you will be in shaded conditions, wear a long-sleeved shirt or bring a jacket.
10. Wear shoes (such as old tennis shoes) and socks if you will not be using waders or high rubber boots. Do not wear sandals or open-toed shoes.
11. Wear protective gloves if handling soil, water, or other materials you believe may be contaminated. After you have completed your field work, wash your hands with soap and potable water.
12. Make sure you know the location of the first-aid kit. If you have any outdoor allergies, bring along any special medication that may be necessary in case of an allergy attack.
13. Wear insect repellent, sunscreen, and sunglasses if the conditions warrant.

14. Never engage in horseplay or distracting activities. Behavior detrimental to the field study will not be tolerated.
15. Do not take chances. Do not threaten the safety of anyone on the field study, including yourself.

Safety Contract

I have read and understand all the field safety rules and I agree to follow these rules. If I fail to follow these rules, I will not be allowed to do the activities.

____

Student's Signature

Name

STUDENT INFORMATION 6.2

Field Study in the Geographic Process

In geographic study, field study often provides the link between background learning and better understanding of a particular area or issue. The field study that you will do in this lesson gives you opportunities to apply what you have learned. You will also collect information to use in the culminating unit activity in Lesson 7. Here is an overview of the general connections to keep in mind during your field study.

DRIFTWOOD

"Fieldwork has an ancient and honorable tradition in Geography. The best way to get to know a place is to see it first-hand, and no one can claim that he truly knows an area until he has seen it for himself."

From *Foreword* by John Fraser Hart in *Introduction to Geographic Field Methods and Techniques*

Applying Geography's Themes to a Field Study

Geography's themes can help you organize information. In your field study, you will use the first four themes (which you have studied in the preceding lessons) as tools for understanding the quality and context of your local river or stream. Later, in Lesson 7, you will pull together this information into a regional perspective on this area, a reflection on the fifth theme, region. Because the information presented in earlier lessons is cumulative and is needed in this lesson and the next one, review that information now. During your field study, keep in mind the five themes, presented here again in brief form:

1. **Location** describes where a particular phenomenon is. Latitude and longitude are commonly used to describe absolute location. Relative location is a position on Earth that is indicated with reference to some other phenomenon.
2. **Place** is a description of the physical and human features found at a particular location.
3. **Movement** investigates spatial interaction—the connections among phenomena, both human and physical, on Earth's surface.
4. **Human-Environmental Interaction** focuses on challenges posed by Earth's physical features and human modification of these features.
5. **Region** is an area on Earth's surface differentiated from surrounding areas by one or more specific characteristics.

(For a more complete review of the themes, look again at the National Geographic Society poster, *Maps, The Landscape, and Fundamental Themes of Geography.*)

Following Analysis With Synthesis

Student Activity 6.3: A River or Stream Field Study marks two important shifts in focus from your previous studies in this unit. First, you will be looking at the real environment rather than reading about it or performing classroom activities. Written text materials and maps are only symbolic representations of the real world. Your main task at the field site will be to recognize and label geographic concepts as they exist in the real world and to look for relationships among the concepts.

Second, your focus will change from **analysis,** the separation of a phenomenon into its component parts, to **synthesis,** the combination of elements to arrive at an overall picture. In Lessons 1 through 5, you were asked to analyze information—to break information down into smaller pieces to better understand it. In this lesson, you will use both methods.

The field study requires you to analyze the phenomena you observe, but you also will make observations about how the whole river system fits together. Because the field study is limited in both time and space, your efforts at synthesis in this lesson will take place primarily in the observations you make at the study site. You will use the process of synthesis again in the next lesson, to study phenomena in a region that you will define.

DRIFTWOOD

"'Contrariwise,' continued Tweedledee, 'If it was so, it might be; and if it were so, it would be; but as it isn't, it ain't. That's logic.'" From *Alice's Adventures in Wonderland* by Lewis Carroll

Selecting a Special River or Stream Topic

As every location is different, every river is different. Though you certainly recognize universal characteristics that all rivers possess, as part of your field study you will also focus on the unique set of physical and human elements that exists at one particular place. In Lesson 7, you will use those observations to describe how your river is unique, how it has a **personality of place** all its own.

So, as part of your field study, you will identify and make observations pertinent to several topics in which you are interested that relate specifically to your river or stream, or to the land nearby. For instance, you may select a particular land use (such as a park, wetland, or a commercial, historic, industrial or residential part of the city on the waterfront). You may choose an activity, (such as boating, fishing, hiking, or camping.) Consider what perspective you might want to take, such as historical, economic, ecological, cultural, or spatial. Perhaps you think a particular land use or activity should be changed or restricted, or expanded.

On your field site visit, you will jot down notes that relate to those topics. Then, in Lesson 7, you will select a single topic and define the region it occupies.

Questions

Write your answers on a separate piece of paper. Use complete sentences.

1. In your own words, briefly define the five themes of geography.
2. What four themes of geography have you studied so far? On which theme will Lesson 7 focus?
3. How does the focus of this lesson differ from that of the preceding lessons?
4. How does analysis differ from synthesis?
5. List several topics of interest to you that relate specifically to your river or stream. Describe the area where you can observe each of these. (This may be your field-study site).

DRIFTWOOD

"Come forth into the light of things,

Let Nature be your teacher."

From *The Tables Turned* by William Wordsworth

Name

STUDENT ACTIVITY 6.3

River or Stream Field Study

Purpose

To identify in a real environment the geographic concepts learned in the classroom.

Background

During this activity, you will study a particular river or stream field site on a topographic map, prepare for a site visit, make observations at the field site, and, finally, analyze and draw conclusions based on that information.

Materials

Per student

- photocopies of writings about rivers and river-related activities
- Student Information 6.1: Safety Guidelines and Contract, signed
- Student Information 6.2: Preparation for Field Study
- permission form (signed by a parent or guardian)
- ruler
- journal
- notebook
- clipboard or other firm writing surface
- pencil
- several sheets of lined and unlined paper
- camera and film (optional)
- appropriate clothing (see Student Information 6.1)
- insect repellent, sunscreen, medication, sunglasses as appropriate
- bottle of drinking water

Per two students

- laminated topographic map of the field site (7.5-minute series; 1:24,000 scale is preferable) (or one photocopy per student of field-site portion of topographic map)

Per group

- Celsius thermometer, alcohol-filled with metal jacket, or electronic
- watch

Per class

- other maps and materials, such as wetlands maps, plat books, EPA reports, conservation department information
- compass
- first-aid kits
- life jackets and tow lines for deeper-water sites
- wading boots for shallower-water sites
- protective gloves
- bottled water for washing
- soap and towels
- paper cups
- large plastic bags for collecting waste

Safety

Follow all the field safety procedures presented in Student Information 6.1: Safety Guidelines and Contract. Do not take chances with safety. Ask questions.

Procedure

PART A. Before the Field Study

1. Review the Safety Guidelines. Sign the Safety Contract and give it to your teacher.
2. Have a parent or guardian sign your permission form; give the signed form to your teacher.
3. Read the entire activity.
4. Carefully examine a topographic map of the study site and assess the features and relationships you expect to observe at the site.
5. On a separate piece of lined paper, create the labels and entries listed in the Observations section of this sheet.
6. Determine the absolute location of your field site to the nearest degree; record it in your own Observations section.

7. Determine and record the location of the field site relative to your school, using cardinal or intermediate compass directions, and record in your Observations section.
8. Calculate and record the distance (straight line and by road) of the study site from your school.
9. On the day of the field study, double-check the materials list and safety guidelines.

PART B. During the Field Study

10. At the field site, complete the Observations section for this activity in your *Rivers Geography* notebook. Be sure you record sufficient information to answer the Critical Thinking Questions. Your answers to the questions will form the basis for classroom discussion following the field study.
11. On a piece of unlined paper, sketch in pencil the river or stream at the study site. Label the natural and human features of the area. Include any of the following as appropriate:
    a. Natural features: estimated width of river (or stream) and floodplain, islands, rapids, sand or gravel bars, oxbows, wetlands, tributaries, springs, obstructions in river or stream channel, vegetation cover.
    b. Human features: human features and human modifications of the river or stream, such as houses and other buildings, roads, railroads, bridges, power lines, dams, boat ramps, riprap.
12. Compare your topographic map with the landscape you see at the study site. Record on your sketch map or on your field trip observations any omissions, corrections, or additions.
13. Add to your record any other observations you feel are worth noting, particularly those relating to the possible topics you have selected for possible further development in Lesson 7. Keep in mind the themes of geography.
14. If you wish, make sketches or take photographs to support the development of your possible topic or topics of particular interest.
15. In your journal, write your impressions of the river or stream. What is its "personality?" What do you feel like when you are at the site? Record all your impressions, including your feelings about the river or stream. Remember that you will use this information in Lesson 7.
16. As necessary, read the remaining questions on this sheet and take pertinent notes at the field site.

PART C. Synthesizing the Field Study

17. After the field site visit, before the next class session, complete on a separate piece of paper the sections labeled Analyses and Conclusions and Critical Thinking Questions.
18. Using the notes on your impressions of the river, compose a journal entry to describe the personality of your river.

Observations

1. Site Identification
   River or Stream Name: ________________
   Location: Absolute (to nearest whole minute) _________
   Relative to school (using cardinal or intermediate compass directions)
   ____
   Distance from school: straight line: _______ km (_____ miles)
   by road: _______ km (_____ miles)
2. Conditions at Site
   Temperature: ____________ Date: _______________ Time: ___________
   Cloud cover: _____________________ Weather: ______________________
3. General description of the site:
4. Observable human uses and activities or evidence of such (List types and numbers):
5. Observable examples of pollution in the river or stream and in its vicinity:
6. Other observations worth noting, particularly those relating to topics for possible further development in Lesson 7:

Analyses and Conclusions

1. What information about the river or stream site can you determine from the field study but not from the topographic map?
2. What possible conflicts exist among various users of the river or stream?

Critical Thinking Questions

1. How do the natural and cultural features of this location relate to features in other locations?
2. What combination of characteristics make this location unique?
3. How have the river (or stream) and its immediate environment changed in form and in function?
4. What positive results can you attribute to changes in the river or stream and its environment? What negative results?
5. How has the significance of the location changed over time? In what ways have local or regional changes altered the relative location of the river or stream?
6. How do conflicts over the use of the river or stream arise? Do conflicting uses lead to misuse? How can differences be resolved?
7. How has this field study changed your impressions of the river or stream? What are your strongest positive and negative impressions?

Keeping Your Journal

1. Compare learning about the river in the classroom to learning about it when the river (or stream) is the classroom. In what ways does one approach complete or fill in information that the other lacks?

How Is Your River or Stream Location Unique?

Focus: Region

The regional theme developed in this lesson serves to culminate the *Rivers Geography* unit. Students describe the unique character of the study region by constructing a brochure to promote the region.

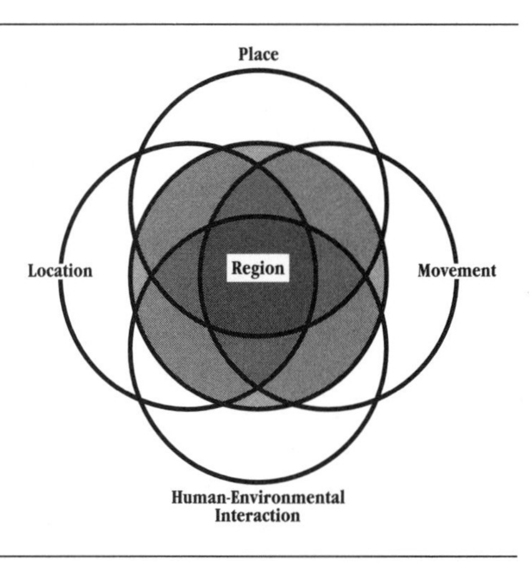

Perspectives and Geography Standards

Perspectives: Spatial; others, depending on topics selected for projects

Geography Standards Receiving Primary Emphasis

This lesson primarily promotes mastery of the following standards:

3. How to analyze the spatial organization of people, places, and environments on Earth's surface
5. That people create regions to interpret Earth's complexity
18. How to apply geography to interpret the present and plan for the future

Geography Standards Receiving Secondary Emphasis

Depending on students' topics for brochures, other standards will also receive emphasis. In particular, students may gain insight regarding the influence of culture and experience on people's perceptions of regions (Standard 6).

Learner Outcomes

Students will:

1. Evaluate previous lessons to make generalizations about the river or stream study site.
2. Explain how relationships at the study site reveal a distinct region.
3. Prepare a brochure promoting the river or stream for some selected purpose.

Time

Five to six class periods of 40–50 minutes per period. (Lesson will require less class time if students complete the brochures as homework). Allow extra time at the end of this lesson if students will do multiple unit-assessment activities.

BEFORE STARTING:	Assign Student Information 7.1: The Region
DAY 1:	Review Lessons 1 through 6 Review Student Information 7.1: The Region Distribute Student Assessment 7.2: River or Stream Brochure and Student Activity 7.3: Evaluating Your River or Stream Brochure
DAY 2:	Review Student Assessment 7.2, and complete Part A, Selecting a Topic
DAY 3:	Complete Student Assessment 7.2, Part B, Creating Your River or Stream Brochure. (Allow extra day if desired.)
DAY 4:	Present brochures Collect brochures, written work for Student Assessment 7.2, and Student Activity 7.3
DAY 5:	Have students perform unit-end assessments, detailed in Appendix A

Advance Preparation

Review Lessons 1 through 6, and read Lesson 7. Prepare to distribute to each student a copy of Student Information 7.1, Student Assessment 7.2, Student Activity 7.3, and any unit-end assessments selected (see Appendix A).

Obtain all necessary materials for this lesson. Finish collecting good brochures, such as those from travel agencies, government agencies (conservation, soil and water, forestry), and the local chamber of commerce, that students can use as models. Make photocopies of field-study site photographs, or, if your computer lab has the capacity to scan photos, have the photographs scanned.

If you want students to use computers in constructing their brochures, schedule student access to computers and familiarize yourself with software programs students will use.

Materials

Student Assessment 7.2: River or Stream Brochure

Per student

scissors

ruler

glue or glue sticks

Per pair of students

poster entitled *Maps, The Landscape, and Fundamental Themes of Geography* (from National Geographic Society)

Per class
maps, photos, slides, and other visuals of regions
model brochures
magazines from which pictures can be clipped
a variety of paper for finished brochures
photographs of the river or stream, if available (from field study or some other source)
Optional
computers
word-processing or desktop-publishing software

Vocabulary

functional region
regional study
systematic (topical) study
uniform region

Background for the Teacher

For most students, the region they identify as the locale for their topic may be the field-study site. For instance, such a regional identification may arise if the field-study site is a county park, adjoins a particular farm or industrial site, or features the town waterfront. Occasionally, student topics may encompass a region outside the field-study site, such as a flyway for particular waterfowl or a stretch of the river often used for recreational boating. Reassure students that their brochures will work well either way, and that, in fact, the purpose is not to define a different region but to explore the concept of region.

For most students, also, the brochures work best if they deal primarily with the particular topic, only secondarily with the concept of region. Again, make sure students know that region is not the dominant criterion for the brochure project, but rather one of the five themes of geography that their brochure should reflect.

Almost any field-study site presents opportunities for significant diversity among student topics. For example, if the study site features a wetlands area, student topics might include:

- Physical: vegetation, precipitation, frequency of flooding, habitats provided, drainage pattern
- Human: construction of buildings or roads; pollutants from industrial, municipal, mining, or agricultural activities; litter; road salt; removal of vegetation; diversion or withdrawal of water from the river. The effects of proposed developments—such as planned residential development, channelization of the river, damming, or construction of a visitor center and observation footpaths—would make exciting brochure topics.

On the other hand, if the field-study site (or a student's preference) featured a historic district on the waterfront, selected topics might emphasize historic settlement, ethnic and cultural studies, architectural comparisons, or significance of the site to the city.

Introducing the Lesson

1. Assign students to read and answer questions on Student Information 7.1: The Region as homework before the first class for Lesson 7.
2. Give students a brief overview of Lesson 7. Review Student Information 7.1. As appropriate, relate this information to your local field-study site.
3. Review discussion of analysis and synthesis included in Student Information 6.2. Make sure students understand that this culminating lesson, the study of region, focuses on synthesis, in which students use all the phenomena studied thus far to create a composite picture of the selected region.
4. Review Lessons 1 through 6, emphasizing how the five themes of geography relate to defining your local region and to your students' selection of topic for a brochure that focuses on a specific aspect of the region.
5. Present examples of single-topic regions as defined by a legally defined boundaries (such as a county), forests, predominance of a particular crop (such as the corn belt), climate, or population density. Contrast single-topic regions with multiple-topic regions, such as metropolitan areas, culture areas, or areas that are defined by the relationship between climate and vegetation.
6. Assign students to read Student Assessment 7.2: River or Stream Brochure and Student Activity 7.3: Evaluating Your River or Stream Brochure for the next class period.

Developing the Lesson

1. Review Student Assessment 7.2. Clarify procedures and expectations.
2. Ask students to share the journal entries from Student Activity 6.3: River or Stream Field Study, in which they described the personality of the river or stream at the study site. Point out that they may use these entries, with modifications, as the basis for the description of the region in the brochures they will create. If students are reluctant to read their entire journal entries, ask them to select single statements from their journals or make general observations based on their entries.
3. Have students discuss the major characteristics of the study region. As students mention characteristics, you may list them on the board.
4. Have students complete Student Assessment 7.2, Part A: Selecting a Topic. As appropriate, review student choices of brochure topics.
5. Display the model brochures so students can examine them in detail. Encourage students to discuss which features make the brochures more effective (and less effective).
6. Have students complete Part B of Student Assessment 7.2, River or Stream Brochure. Determine the amount of time you allow students for their brochures (in class and as homework) on the basis of student skills, available resources, and the quality of finished product you (and the students) desire.

Concluding the Lesson

1. Have students present their brochures to the class. You may generate additional interest by inviting parents, administrators, or other classes to the presentations. Display completed brochures in the classroom or a more public (but protected) display area, such as the school library or public library.
2. Review Critical Thinking Questions in Student Assessment 7.2.

Assessing the Lesson

1. Have students turn in Student Activity 7.3: Evaluating Your River or Stream Brochure with their brochures.
2. Perform unit-end assessment. From among the following unit-end assessments in Appendix A, select the activities that best suit your preference and circumstances:
   - student portfolios of best and most significant work from *Rivers Geography*
   - forced-choice unit test
   - one or two performance-based assessments emphasizing demonstration of geographic skills
   - telecommunications project in which students download and analyze river data provided by Rivers Project students in numerous academic disciplines and locations
3. Collect and assess student journals. Here is a suggested scoring rubric, which you can use with the performance criteria included in the student handout:

Scoring Rubric

Score	Expectations
0	No response attempted.
1	Work shows little or no understanding of geographic concepts, lacks attention to detail, and shows little or no application to real-life situations.
2	Work shows limited understanding of geographic concepts, limited attention to detail, and limited application to real-life situations.
3	Work shows generally satisfactory understanding of geographic concepts, includes some detail, and demonstrates attempt to apply concepts to real world, although the application may be incomplete.
4	Work shows strong understanding of geographic concepts, contains valid applications of the concepts to the real world, and is supported with detailed observations.

Extending the Lesson

1. As an alternative project, have students prepare a one-page newspaper display about their river or stream, such as might appear in a feature section of the local newspaper. For specific teaching activities on feature writing, see *Rivers Language Arts*.
2. In lieu of a brochure, allow some students to create a short videotape presentation about an issue relating to their river or stream.

Name

The Region

A region is an area distinguishable from other surrounding areas by some unifying feature. A region is a geographic tool that is helpful in selecting a portion of the world for closer study.

The bounds of a particular region depend on what criteria the investigator has in mind. Regions may exist in a great variety of forms and serve many purposes. A region's distinguishing feature may be legal status (such as county boundaries), a dimension of the landscape (such as common watershed), use (such as irrigated farmland, industrial development), cultural (concentration of people of a particular ethnic group), a historic district, distinct architectural types, or a certain educational philosophy.

Any particular location is part of a multitude of regions. Your school campus may be part of the city or town where you live; it is also part of larger political regions, such as a county, state, and nation. It is also included in particular climatic and landform regions, which do not conform to political boundaries. It is also located in the watershed drained by a particular river, perhaps your project river. All these are regions, just based on different topics.

Regions may be based on one topic or many. Sometimes investigators must quantify a feature in order to determine regional boundaries. "Corn Belt" may refer to the area in which corn is an important crop. One study, however, might use the relative value of corn to other crops, while another uses the ratio of acres planted in corn per square mile.

Scale may likewise be highly variable. One region may overlap another. Every region, however, must satisfy two requirements: it must occupy an area on the earth's surface, and it must have at least one unifying feature that distinguishes it from surrounding areas.

A geographic study may be either systematic or regional, depending on where the geographer begins the investigation. In a **systematic** or **topical study,** the geographer starts by selecting a topic or subject. He or she then attempts to identify and understand in which regions the topic appears, called the **spatial distribution** of the topic.

In a **regional study,** the geographer begins by identifying a region, an area in some way different from other nearby areas. Then the geographer collects information about topics affecting that area in order to better understand the region. Regional studies are synthetic, because the geographer uses a variety of topical information to create an overall picture of a particular region.

DRIFTWOOD

"The method of scientific investigation is nothing but the expression of the necessary mode of working of the human mind."

From *Our Knowledge of the Causes of the Phenomena of Organic Nature* by Thomas Henry Huxley

DRIFTWOOD

"'We face a real dilly,' Schneider was saying. 'Because Colorado is such a popular state, fifty thousand newcomers want to move in each year. We'd like to welcome them, but we haven't enough water. And within the state itself, twenty thousand of our rural people a year want to move into Denver. Love to have them, but no water. We also have scores of industries that want to establish their headquarters here. Executives want instant skiing, and we need their tax dollars. But we simply don't have the water."

From *Centennial* by James A. Michener

Despite their different starting points, the two types of geographic studies—systematic and regional—may produce similar conclusions. For example, consider two hypothetical studies. In the first study, a systematic study on the subject of coffee, the investigator might explain where coffee was first domesticated (Ethiopia), the diffusion of coffee to other areas, the importance of coffee as an economic commodity, and the physical and cultural conditions necessary for the production of coffee. Each of these topics requires the identification of regions, including the very important coffee-producing region in the Brazilian Highlands near the city of São Paulo.

In the second study, a regional study, the investigator focuses on the region of Brazil. The investigator discusses the topics most important in distinguishing Brazil as a region, such as landforms, climate, population distribution, and economics. Because coffee production is an important component of the Brazilian economy, the regional study would include the topic of coffee production in the São Paulo region. Despite what appear to be two very different methods of study, the systematic and regional approaches are complementary. Each is a necessary part of the other.

Usually regions are not like puzzle pieces that fit neatly together. They usually overlap like leaves on a forest floor. Regions that are part of a formally defined system, however, such as legally defined states, counties, or school districts, do have neat divisions.

A **uniform region** displays some significant similarity throughout its area in terms of the defining topic. A **functional region** has a focal point that is the center of some regionwide activity. Your school district can be seen as a uniform region in that it is legally fixed, exists to educate students who live within its boundaries, and is supported by taxes collected from owners of property within the district. The school itself may be viewed as the center of a functional region, because it is the focus of spatial interaction, such as student arrivals and departures.

DRIFTWOOD

You can think of regions as being like the week's laundry scattered on the laundry room floor. The various sizes, shapes, and categories of regions are represented by overlapping sheets, T-shirts, towels, and socks.

DRIFTWOOD

"It is not enough to have a good mind. The main thing is to use it well."

From *Le Discours de la Methodé* by René Descartes

Questions

Write your answers on a separate piece of paper. Use complete sentences.

1. In your own words, explain what a region is and give two examples.
2. Compare and contrast a uniform region with a functional region.

Name

River or Stream Brochure

Purpose

To prepare a brochure that describes the geography of a local river or stream topic.

Background

In this activity, you will develop a topic related to your local river or stream and describe the area in which it occurs.

Brochure Topic and Region

When you make your topic selection, choose a subject on which you to want to inform people: a potential human activity (canoeing, fishing, bird-watching); some problem (increased sedimentation, pollution, misuse, the need for habitat preservation); the historic or economic roles of the river; or particular ecosystems or natural characteristics.

In defining the region of study, remember that you must be able to distinguish it from the areas around it. Do not be overly concerned about defining your region precisely, however. Instead, concentrate on showing how your topic interacts with various physical and human features there to produce a unique "personality."

You do not need to define a large area for this project. Instead, the topic and region should be small and manageable enough that you can observe them first-hand.

For instance, if your topic of interest is wildlife, you might select a particular wildlife refuge or park as your region. If you're interested in water recreation, you may select an area commonly used for boating, swimming, or fishing. If your concern is water pollution, then your region may be the area impacted by municipal or industrial wastes, or by particular agricultural practices that contribute to sedimentation or chemical runoff. You also may decide to focus on your community and its relationship to the river. In many cases, the region for your topic may be your field-study area (though you will still need to define it as a region).

Materials

Per student
- scissors
- ruler
- glue or glue sticks

Per pair of students
- poster entitled *Maps, The Landscape, and Fundamental Themes of Geography* (from National Geographic Society)

Per class
- maps, photos, slides, and other visuals of regions
- model brochures
- magazines from which pictures can be clipped
- a variety of paper for finished brochures
- photographs of the river or stream, if available (from field study or some other source)

Optional
- computers
- word-processing or desktop-publishing software

Brochure Contents

Once you have selected your topic and defined the region it occupies, you'll prepare a brochure promoting or describing that region in terms of that topic. A brochure is a brief essay in pamphlet form. It is an arrangement of factual information that highlights a place, an idea, or an activity. In addition to written information, a brochure may include photographs, drawings, or maps. Because a brochure has limited space, you must select the contents of your brochure carefully in order to achieve maximum impact.

Though your brochure must be factual, you are encouraged to describe the personality of the river in creative terms. Pay attention to personality of place, the combination of cultural and natural features that create a unique character.

Because the brochure must be factual, you should review all the information that you have learned and collected during the course of this project. You also can supplement that information with further research.

In Lesson 6, you used the five themes of geography to organize and make observations during your field study. In this lesson, you will use them to plan the structure and content of your brochure, to help you achieve some synthesis of issues relating to your topic and region.

So, identify and use the five themes of geography as part of your process of selecting and exploring your study of your local region. To select the relevant characteristics of your region, use place characteristics and the interaction of humans with the physical environment. To help the reader identify the area you describe, give its relative location. To highlight the uniqueness of your region, explain the interrelationships in the area. Illustrations of movement in your region might include identifying the natural movements of its wildlife inhabitants, its hydrologic cycle, or the activities or proposed activities of humans.

Procedure

PART A. Selecting a Brochure Topic

1. Review Lessons 1 through 6, focusing on how the five themes of geography relate to your potential topics. Read Student Information 7.1 to review information about regions.
2. Finalize your selection of a topic that relates to your local river or stream. Although you may have several messages about the river or stream you want to communicate, your brochure will be most effective if you focus primarily on one topic. Make sure the topic is narrow enough that you can give it a thorough treatment.
3. Create a list of ways to define the area that relates to your local river or stream (such as watershed boundaries, legal jurisdictions, water use, a particular wildlife habitat, or river feature).
4. From among these regional definitions, select one that fits well with presenting your topic in a brochure. Select an area you can define on the basis of one characteristic.
5. Select the most significant features and relationships of the region as you have defined it.
6. Complete the Analyses and Conclusions segments of this activity.

PART B. Creating Your River or Stream Brochure

7. If your teacher has provided examples of brochures, look at them critically.
8. Research the topic and collect information. Use Student Activity 7.3: Evaluating Your River or Stream Brochure as a guide for questions to try to answer. If time allows, you may want to look in library records of local newspapers or other relevant materials.
9. Plan visuals for your brochure, such as sketches or photographs available from the field-site visit. Cut out or photocopy appropriate photographs from magazines. Photocopy other visual materials that you might like to include in your brochure. Copy or create maps as appropriate.
10. Decide precisely what you want to say. Jot down notes about each point, then polish the statements you intend to use. Revise your written passages until the messages are clear and fit together to make the impression you want.
11. Decide on the size and format of your brochure. For example, if you are using a letter-sized sheet of paper, do you want to fold it into four or six panels?

12. Make a mock-up (rough draft) of your brochure and evaluate it critically. Make sure your messages fit on the paper. Plan where you want to place photographs, maps, or drawings. As necessary, have visuals photocopied in a reduced size to fit in your brochure. If you have access to a scanner and a computer, you can scan the photographs into the computer and use the computer to adjust the size. Is the brochure clear? Are the photographs and art appropriate? Is the message apparent? Check the guidelines again. Identify and make any necessary revisions.
13. Select the paper you want to use and prepare the final version of your brochure. A neatly printed brochure will be acceptable, but you can achieve a more professional appearance if you prepare the text on a typewriter or word processor and paste it on the brochure. If you want, use a computer software program to generate the entire brochure.
14. Do Critical Thinking Questions and Keeping Your Journal.
15. Complete the self-evaluation form (Student Activity 7.3) and hand it in with your brochure.

Analyses and Conclusions

Write your answers on a separate piece of paper. Use complete sentences.

1. Describe the topic you have selected for your brochure. Explain why you selected this topic.
2. Define the region within which you will focus your exploration of this topic.
3. What concepts from Lesson 1 through 6 do you plan to include in your brochure? Give a reason for including each.
4. What are the most significant features and relationships of the region as you have defined it?

Critical Thinking Questions

Write your answers on a separate piece of paper. Use complete sentences.

1. Compare your brochure with the collage you created in Lesson 1. What are the advantages of each medium for communicating ideas?
2. Which special-interest groups would be most likely to agree with the ideas expressed in your brochure? Which special-interest groups would disagree? Why?
3. In what ways is studying a topic in a particular region more useful than studying that topic worldwide?

Keeping Your Journal

1. How have your impressions of rivers and streams changed during the *Rivers Geography* unit? Do you have stronger opinions about topics related to rivers and streams? Which topics?

2. What topics related to rivers and streams would you like to know more about? How will you satisfy your need to know more?
3. Does the personality of your local river or stream differ from that of others you have visited? How?
4. What do you like best about the local river or stream? What do you like least?
5. What would help others appreciate the river or stream?
6. What changes to the local water region do you foresee?

Name

Evaluating Your River or Stream Brochure

The following questions will be the basis for evaluating your brochure. In the "Student" column, score yourself in each category, based on how well you believe you accomplished each task. Award points using the following scale:

0 = Task not accomplished
1 = Needs improvement
2 = Generally satisfactory performance
3 = Excellent job

If directed by your teacher, hand in this form with your brochure. You may place additional comments on the back of this sheet.

	Evaluator	
	Student	**Teacher**
Does the title of the brochure indicate its purpose?	___	___
Does the introduction describe the brochure's purpose?	___	___
Does the brochure describe the location of the river or stream?	___	___
Does the brochure describe the natural and human characteristics of the river region?	___	___
Does the brochure explain how humans have modified the natural environment?	___	___
Does the brochure include patterns of circulation and connection with other regions?	___	___
Does the brochure explain how the region is related to other regions?	___	___
Does the brochure explain how the region is unique?	___	___
Does the brochure use photographs, maps, charts, or drawings effectively to emphasize and clarify its purpose?	___	___
Does the brochure have design and language appealing enough to make someone want to read it?	___	___
Does the brochure present the message you intended?	___	___
Total Points (33)	___	___

APPENDIX

A

Unit-End Assessments

This appendix contains the unit-end assessments for *Rivers Geography*. (Assessment activities for specific lessons are contained within those lessons.) As is true throughout this unit, this appendix contains two types of assessment—performance-based and forced-choice. From the assessment tools presented in *Rivers Geography*, utilize what suits your class, curriculum, and preferences. Feel free to use your own assessment tools, performance criteria, and scoring rubric.

Teacher Notes for Portfolio (page 199)

Focus

Each student selects and arranges multiple assessment items that demonstrate the student's learning of scientific content, attitude, and skills.

Advanced Preparation

Look through one or more of the lessons in *Rivers Geography*. In each lesson, a Teacher Notes section gives suggestions on activities and projects for assessing and extending the lesson. As you select which assignments you will give, make sure students will have a variety of opportunities to express their abilities and interests during this study of river geography.

Based on your assignment plans, decide what kinds of items students may (or must) include in their portfolios. Modify the accompanying handout, Student Assessment: Portfolio, as appropriate to reflect your approach.

If you have used portfolio assessments for special projects, collect good examples to share with your students. If not, select some ideas from different lessons to make a sample portfolio.

You may have students receive the handout on portfolios either at the beginning of this unit or as you near the end. If you describe the portfolio project or share the handout early, you may have students collect, as the *Rivers Geography* unit progresses, their choices of what they plan to include in their portfolios. If so, you might have them keep their selections in manila folders in the classroom. To conserve space, you may limit the total items per folder to five at any time.

Materials

Per student

Manila folder

Plastic report cover or three-ring notebook

Developing the Lesson

1. Have students read the handout, Student Assessment: Portfolio. Illustrate the diversity of work students may select for their portfolio by showing a sample portfolio that includes a variety of items such as special reports, maps and sketches, other writings, interviews, and group reports.
2. If relevant, show students where in the classroom they should keep their portfolio selections during the *Rivers Geography* project.
3. Show how students should prepare the portfolio for final submission, such as in a bound notebook, folder, or a three-ring notebook.

Scoring Rubric

Score	Expectations
0	Portfolio is absent, or only random papers are included, with no summary statements written.
1	Portfolio is incomplete. Some required sections are absent; Introduction, Reflections, and Conclusions sections are poorly written or fail to address suggested components.
2	Portfolio is generally satisfactory. Materials are properly assembled and show some variety; Introduction, Reflections, and Conclusions sections provide basic information about the project and student involvement
3	Portfolio is excellent. Materials display a variety of skills and accomplishments; Introduction, Reflections, and Conclusions sections are thorough and well-presented.

Teacher Notes for Creating a Geographic Presentation (pages 200–201)

The performance-based assessment based on scenarios, Creating a Geographic Presentation, directs the student to assume the role of a geographic consultant to organize studies about the impact of proposed human changes to the physical (and human) environment. This assessment is based on the five geographic skills described briefly in the introduction to *Rivers Geography* (and developed more fully in *Geography for Life: National Geography Standards 1994,* referenced in Appendix C.): asking geographic questions, acquiring geographic information, organizing geographic information, analyzing geographic information, and answering geographic questions.

You may let students choose which scenario to work with, tell them which scenario to do, or develop your own. You may even want to create a scenario based on an actual local issue. You may adjust the complexity of the task and end product to suit student skills and time available.

The student handout contains performance criteria for students to use. Direct students to pay primary attention to meeting the performance criteria listed (which emphasize the geography skills and the concepts developed during *Rivers Geography*), rather than to the determination of an answer to the problem posed by the specific scenario.

Scoring Rubric

Score	Expectations
0	No response was attempted.
1	Demonstrates little or no understanding of concepts, and fails to identify important elements. Shows minimal evidence of a solution process, and arrives at an unclear solution. Applies an inappropriate strategy.
2	Demonstrates some understanding of concepts, and identifies some important elements, but fails to understand relationships. Shows some evidence of a solution process, although the process is difficult to identify. Uses an appropriate strategy, but the application is unclear.
3	Demonstrates general understanding of concepts, and identifies most of the important elements of the problem. Shows a general understanding of relationships. Uses a satisfactory solution process with minor omissions. Applies an appropriate strategy that is clearly presented.
4	Demonstrates complete understanding of concepts. Identifies all the important elements of the problem and explains relationships. Completes the solution process by systematically applying an appropriate strategy.

Teacher Notes for Rivers Geography Assessment (Forced-Choice Test) (pages 202–206)

The unit-end forced-choice test is of the multiple-choice type. The test is designed to measure student learning about concepts relating to geography's five themes. The answers assess knowledge, understanding, and application. The following chart shows how each of the 30 test questions relates to the themes and skills of geography.

	Knowledge	Understanding	Application
Location	1, 2	11, 12	21, 22
Place	3, 4	13, 14	23, 24
Human-Environmental Interaction	5, 6	15, 16	25, 26
Movement	7, 8	17, 18	27, 28
Region	9, 10	19, 20	29, 30

Test Answers

1. C	11. A	21. B
2. C	12. C	22. B
3. C	13. B	23. D
4. B	14. D	24. B
5. B	15. D	25. A
6. A	16. A	26. C
7. A	17. A	27. A
8. B	18. C	28. B
9. A	19. A	29. C
10. B	20. D	30. B

Teacher Notes for Challenge Project: Telecommunications (pages 207–208)

Focus

In this activity, students use data from other schools to expand their knowledge of rivers and streams. This activity emphasizes the importance of analyzing collected data and looking at results over time and distance. Once students have retrieved additional river data electronically, they will interpret and report those data. They may also contribute their own material electronically for other students to use.

Time

Three or four class periods of 40–50 minutes per period, with additional homework time for preparing presentation

Advance Preparation

Prepare to supply students with Challenge Project: Telecommunications. Gather all necessary equipment and materials. If your class must share computer capabilities with others in the school, make arrangements for your class to have appropriate access.

Before having students try to connect electronically with the Rivers Project, make these connections yourself, to make sure you can give clear, accurate, and up-to-date instructions on this process. As necessary, create and distribute a handout on making these connections. As necessary, work with your school's computer resource person or other skilled individual to give students proper preparation and guidance for the telecommunications portion of this activity.

In this project, you may have students provide information to the bulletin board, download and analyze information from the bulletin board, or both.

If students will download and analyze information, decide how students will present their investigations, such as written reports, oral presentations, maps, graphs, poster presentations, or a combination of these techniques. Prepare a handout of presentation guidelines if different from what is included on the student sheet. If students will make group presentations, decide how the group and the individuals will be evaluated.

If students have made observations and reports about the field study site or local community during *Rivers Geography* that other students may find useful (such as a streamwalk report or original research about a local issue), suggest that they uplink them onto the bulletin board. (This does not require as much geographic skill or knowledge; you may use it as an optional activity for students who have finished the first part of this project or who have made significant contributions.)

Materials

Per group
computer
telecommunications system
data from a rivers bulletin board or web site
maps of rivers, streams, and watersheds being studied
graph paper
Optional
spreadsheet program for sorting and managing data

Developing the Lesson

1. Have students read and discuss Challenge Project: Telecommunications.
2. Explain that the rivers bulletin board or web site that the students will be using includes the results of observations, water-quality tests, writings, and other data from thousands of students in many geographic areas. Explain the school or classroom computer capabilities and your class access to such facilities. Ascertain which students have experience communicating via e-mail and the World Wide Web.
3. If students will be downloading information, describe the form in which students will present the results of their research, such as written, oral, maps, graphs, poster, and so forth. Distribute handout of presentation guidelines as appropriate, including deadlines.
4. Divide the class into three or more groups to work on the telecommunications portion of this activity. Try to distribute students with computer networking skills evenly among the groups working on this project. Each group should choose a specific river, watershed, or geographic area to investigate. As a group, have them explore the database to see what sites within that watershed or geographic area have data available, and the dates of any relevant data. For specific teaching activities and student information sheets of water-quality tests, see *Rivers Chemistry*.
5. Working with the rest of his or her group, have each student select an area for individual research that fits within the group's general area.
6. Working in groups, have students uplink their information, or retrieve from the rivers bulletin board or web site observations and tests required for each student's individual research endeavor. Peer coaching in using telecommunications should help students with little background in this technology.

7. While some students are working with the computers, have others investigate their test site(s) using maps, atlases, and other library resources.
8. Have students map or graph their data. For specific teaching activities on graphing, see *Rivers Mathematics*.
9. Have students complete their written reports, poster, or other presentation.
10. If students will be providing information onto the Rivers bulletin board or web site, describe how they should prepare such information, and how they will do the transfer (create a handout specific to your computer system as needed.)

Notes

Accessing and donating information via telecommunications may be a new venture for many students. The technology for supporting a database for information changes too quickly to adequately explain in detail all the steps that are needed to access the information.

As was mentioned in the introduction to the Rivers Project curriculum, the Rivers Project includes two telecommunications systems designed for students to use in exchanging data with students at other schools. Information on how to access the Rivers Project via e-mail and on the World Wide Web is included on page x. *Southern Illinois University at Edwardsville (SIUE),* the *Office of Science and Mathematics Education (OSME),* and *Rivers Project* are key words to finding the Rivers Project home page for information on using the database, for asking questions about the project, and for connecting with other schools testing their local river or stream.

You may decide to have all students in the class select rivers and streams within the same watershed as the field site you investigated. If enough data are available, students could even choose sites upstream and downstream from the field site. As students work in groups to select their individual topics, you may require that they coordinate their selection of topics, so they investigate related issues. For instance, projects within each group, when taken as a whole, may involve:

- investigation of streamwalk information for other field study sites along the same river or stream as the field study site, or streamwalks at the same site performed at different times
- comparison of reports about related topics about the same watershed, provided by students at other schools
- inclusion of certain common elements, such as historical background, geology, or weather conditions
- investigation of the nine tests of water quality for the field study site
- investigation of water-quality test results under particular conditions, such as: a specific test under different weather conditions or seasons; a specific test at different sites along one river or stream; a series of tests upstream and downstream from a specific area of interest (such as a city or industrial complex)

You might want students to research their topic or area as background for their findings. This might include historical background on uses of the river and land along the river, a study of the geology along the river, or the typical (or special) weather conditions of the area. Maps of the area with the location of the different sites shown may be required.

Performance Criteria and Scoring Rubric

Because this challenge project can take many forms, performance criteria should fit the specific tasks and activities involved. In most cases, the assessment criteria can be similar to those already provided for other assessments. Criteria for the telecommunications portion of the assessment will depend on student access to and experience with telecommunications systems. For certain activities, you may want specific performance criteria and scoring rubrics. Here are some possibilities.

Written Analysis of Telecommunications Data

Does the student convey knowledge through the writing? Is the written presentation logical, persuasive, and informative? Read the student's entire writing sample. Then reread, using the performance criteria included in Student handout on Creating Geographic Presentations (or your own).

Scoring Rubric

Score	Expectations
0	Nothing is written.
1	Narrative is minimal, has little relevance to geographic themes and standards, or does not describe the work correctly. Maps and other visuals are minimal, missing, or not relevant to the topic. The group relied on others outside their group to retrieve information from the Rivers Project database.
2	Students adequately describe the data they have collected from the database and make some connections between geographic themes and the information collected. Cause-and-effect relationships are usually missing. Some visuals, such as maps, tables, and graphics do aid in communicating the generalizations in the presentation. At least one student in the group was able to manipulate the telecommunications system.
3	Students make meaningful connections between geographic themes and the information collected. They display data clearly in maps, tables, and graphs appropriately, illustrating their major points. Topics are discussed in a logical sequence and information is summarized. All students in the group have accessed the telecommunications system with minimal help.
4	The group fulfills the requirements for a three; has reasonable explanations or correlations between collected information and geographic themes; and has researched some relevant topic

beyond requirements for the course to include in their discussion or otherwise shows interest and scholarship in environmental concerns.

Group Map Presentations

If map presentations are made during a parent night, for an environmental program, or other school function, students may present their results several times to different audiences. An evaluation form may be used for general comments from others about the presentations. If the oral report is for the class, peer review may be included along with your review.

If the map presentation is a group project, it should reflect the different areas investigated by each of the students in the group.

Scoring Rubric

Score	Expectations
0	Nothing is written.
1	The map/presentation shows little preparation or has incorrect or meaningless information.
2	Students have followed the criteria for preparing the map and have some correct and meaningful information. If presentation is a group project, it is the result of efforts of most, if not all, of the group. This score may be given for groups in which one element (either map or presentation) is good, but the other is poor.
3	Both the map and presentation meet most of the criteria. All members of the group participate.
4	Reserve a four for groups that meet all criteria. Exceptional efforts should be shown in either the map or the presentation. All members of the group participate in preparing the map and in the presentation.

Portfolio

Introduction

A portfolio is a collection of work that you deem to be your best and most representative. In preparing your portfolio, you should include material that you would be proud to show others and that you may keep. The purpose of a portfolio is to enable you to see your own progress and to share what you have learned with others. A well thought-out portfolio will allow you, your classmates, and your teachers to assess what you have learned and done.

Procedure

1. During this study of *Rivers Geography,* you will keep some of your best papers in a manila folder in the classroom. You may update the contents of this folder regularly, so that the folder contents represent your best work. These items may include a compilation of entries from your journal, maps and sketches, reports, poems or stories, artwork, articles you have written for a newspaper or published journal, or any other evidence of the effectiveness or benefits of this project.
2. Write an overview of the *Rivers Geography* project. This overview should include a brief introduction, a reflection on what you have learned and done, and an afterword or concluding statement. Address your writing to an adult who has no knowledge of this project but whom you want to impress positively. Here is what to include in these segments.

   In *Introduction,* include information about the project, its purpose, and the major activities in which you and your class participated.

   In *Reflections,* answer the following questions:
   - What new information or skills did you learn or discover while participating in the *Rivers Geography* unit?
   - Which part of the project was most helpful to you? Which part did you most enjoy?
   - About which topics have you developed increased awareness or interest?

   In *Conclusion,* describe how your attitude toward the environment in general and rivers and streams in particular have changed. Explain how these changes may influence your behavior.
3. Complete your portfolio by assembling all materials into a binder, such as a plastic report holder or a three-ring notebook. Place your materials in the following order:
   - Title page, including your name, class, and date
   - Introduction
   - The five items you have selected to represent your work
   - Reflections
   - Conclusion

Materials

Per student

- Manila folder
- Plastic report cover or three-ring notebook

Performance Criteria

The successful portfolio has the following characteristics:

- Materials are assembled as indicated in this assignment sheet.
- Selected materials illustrate a variety of skills and accomplishments.
- Uses correct, standard written English.
- Presents ideas clearly.
- Introduction, Reflections, and Conclusions demonstrate analysis and synthesis of what has been offered in the *Rivers Geography* unit.

Creating a Geographic Presentation

Purpose

To demonstrate your competence in geographic skills.

Background

Assume you are a geographic consultant hired to make an oral or written presentation about your local river or stream. Here are the possible scenarios:

Scenario A: Government agencies and conservation groups are pushing for the restoration of wetlands along your river or stream and its tributaries. Before proceeding, they contact you for advice.

Scenario B: You have been asked to make a presentation about the importance of your local river or stream. It should contain references to both the river or stream as part of a physical system and to its use by humans.

Use the following geographic skills to conduct your project:

- **Ask geographic questions,** such as: Where is it located? Why there? To what else is it related? How will change affect it or other phenomena to which it is related?
- **Acquire geographic information** from a variety of sources, both primary (fieldwork) and secondary (published information). Visit sites, conduct interviews, and do library research.
- **Organize geographic information** systematically by classifying it into logical categories: visual and written, supporting and opposing point of view, or human and physical features. Create graphic displays and map data.
- **Analyze geographic information** by looking for connections and relationships in the information you've gathered. Identify trends, interpret data, and attempt to understand spatial patterns.
- **Answer geographic questions** by developing generalizations and conclusions. Be objective. Finally, prepare your report for presentation.

Procedure

1. From the two scenarios described earlier in this sheet, select one to investigate (or do the one selected by your teacher). Complete the steps following for the selected scenario.

SCENARIO A. Wetlands Restoration

2. Select at least three possible sites for wetland restoration.
3. Identify the individuals and groups the restoration would impact.
4. Determine the suitability of each site for each potentially affected individual or group.
5. Construct a matrix (table) that you can use to summarize and analyze the suitability of each site for each potentially affected individual or group.
6. Write a summary of your findings, and make a recommendation.
7. Assemble all your written elements of this process into a report, with title page, visual support, table of contents, and list of sources.

SCENARIO B. Presentation About Your Local River or Stream

8. Decide on five major points you want to make about your local river or stream.
9. Find a variety of primary and secondary sources to support these points.
10. Based on the evidence you have found in these sources, develop logical generalizations about your river or stream.
11. Construct a written (or oral, if directed by your teacher) presentation, including appropriate visuals. If a written presentation, it should contain a title page, table of contents, and list of sources.

Performance Criteria

The successful geographic report or presentation:

- Uses correct English.
- Uses geographic vocabulary correctly.
- Demonstrates understanding of key geographic concepts and how they relate with one another and with concepts in other academic fields.
- Illustrates relationships of phenomena existing at several scales (local, regional, global).
- Analyzes information from a variety of primary and secondary sources.
- Develops logical generalizations based on combinations of geographic evidence.
- Displays data in maps, tables, and graphs as appropriate that aids in communicating the main points.
- Contains other appropriate visuals to illustrate generalizations.
- Cites sources used.

Rivers Geography

Place the letter indicating the best answer for each question in the space provided. You will need a ruler to answer questions 21 and 30.

_____ 1. The location of a feature compared with the location of a known familiar feature is called
A. absolute location
B. similar location
C. relative location
D. common location

_____ 2. Which of the following is a relative location?
A. 30 degrees S, 45 degrees E
B. a town on a road map with coordinates A-8
C. a city at the mouth of a river
D. a person's street name and house number

_____ 3. Physical features include the following:
A. factories, lakes, soil
B. dams, highways, climate
C. vegetation, animal life, oceans
D. schools, forests, playgrounds

_____ 4. Human characteristics include the following:
A. parking lots, deserts, mountains
B. shoe stores, schools, hospitals
C. waterfalls, roads, streams
D. shrubbery, rainfall, shopping malls

_____ 5. Which of the following is the most likely reason for increased sediment load in a stream?
A. extensive paving of nearby land
B. decreased vegetation cover
C. increased recreational use of the stream
D. increased discharge of industrial waste into the stream

_____ 6. To prevent flooding, humans construct
A. levees
B. locks
C. riprap
D. cut banks

_____ 7. Which of the following refers to the movement of one's ancestors into this country?
A. immigration
B. emigration
C. ethnicity
D. genealogy

_____ 8. Which of the following is *not* a part of the hydrologic cycle?
A. transpiration
B. stratification
C. evaporation
D. runoff

_____ 9. A region
A. is an area that is in some way distinguishable from other areas
B. must have a central focal point where interactions occur
C. is an area that has legally defined boundaries
D. is a particular area that does not overlap other regions

_____ 10. The Mississippi River watershed is an example of a
A. stream stage
B. region
C. tributary
D. river site

_____ 11. Select the group in which each item listed can be used to express location:
A. latitude and longitude, compass direction and distance, reference to a known point
B. compass direction and distance, map scale, latitude and longitude
C. map symbols, contour lines, hydrologic cycle
D. geographic grid, map scale, population of a city

_____ 12. Which of the following would most likely change the relative locations of two cities on opposite sides of a river?
A. dredging the river channel to facilitate water transportation
B. construction of levees and other flood barriers
C. construction of a bridge across the river
D. construction of reservoir dams on the river's tributaries

_____ 13. High human population densities most commonly occur on Earth in
A. deserts
B. plains
C. mountains
D. rain forests

_____ 14. Which is the least likely relationship to be found in a river environment?
A. floodplain and wetlands
B. mountain stream and braided channel
C. low gradient and levees
D. oxbow and braided channel

____ 15. Select the group in which *all* items listed are examples of human modification of the natural environment.
A. riprap, climatic classification, pollution
B. channelization, draining swamps, drawing topographic maps
C. levees, immigration, logging
D. boat ramps, irrigating crops, bridges

____ 16. Select the activity that features human interaction with the physical environment.
A. water-quality testing
B. the relationship of precipitation amounts and vegetation
C. the effects of glaciation on landforms
D. mapping watersheds

____ 17. Which of the following questions would most likely lead to a study of spatial interaction?
A. From where did the first settlers of my community come?
B. How did people alter the physical environment to settle here?
C. What characteristics make the area unique?
D. What misuses of the river can be observed?

____ 18. Which would have the greatest impact on the movement of water in the hydrologic cycle?
A. pollution of streams
B. construction of bridges
C. building reservoirs
D. barge traffic on the river

____ 19. Select the group in which *all* items are examples of regions.
A. Iowa, a watershed, a floodplain protected by a levee
B. a wildlife preserve, a tributary, topography
C. a lock and dam, an area of similar climate, a valley
D. a township, population, a forest

____ 20. Which of the following could change the legally defined boundaries of a levee protection district?
A. a city on the floodplain
B. a flood
C. the desire to remove levees and restore wetlands
D. action by the levee district governing body

Use the map provided to answer Questions 21 through 30.

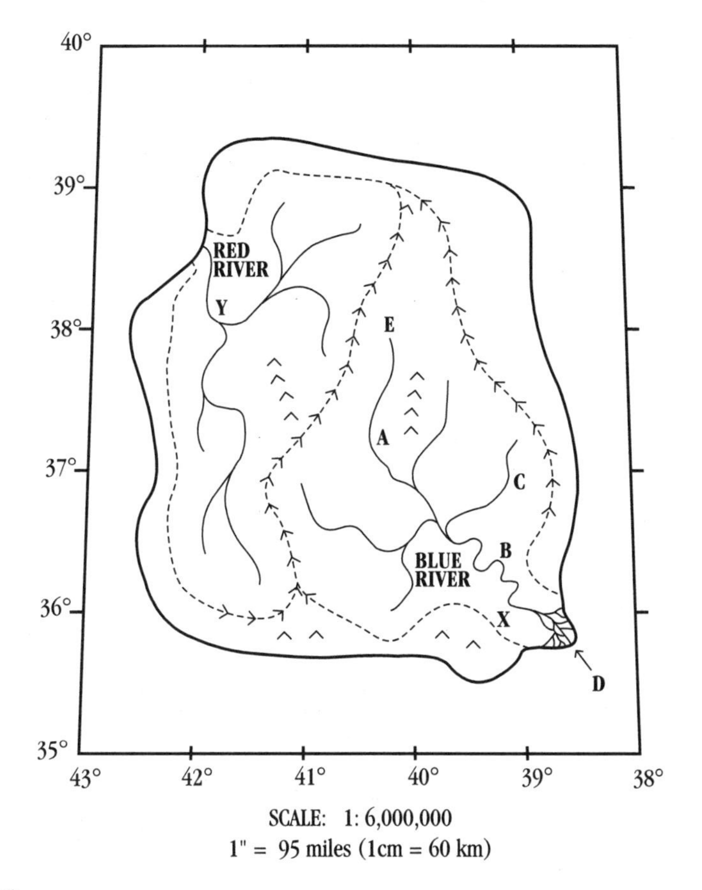

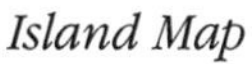

Island Map

____ 21. Which is the best description of the relative location of city Y?

A. near a major river

B. 200 miles NW of city X

C. 38 degrees N, 42 degrees W

D. on the floodplain

____ 22. Which is the best description of the absolute location of the city X?

A. near the mouth of the river

B. 36 degrees N, 39 degrees W

C. 39 degrees N, 36 degrees W

D. 39 degrees S, 36 degrees W

____ 23. At which point is the greatest amount of deposition most likely to be found—A, B, C, or D?

____ 24. The stream labeled C can best be described as a

A. wetland

B. tributary

C. floodplain

D. divide

____ 25. At which point is a hydroelectricity generating station most likely to be built?

A. point A

B. point B

C. point X

D. point D

____ 26. At which point are levees most likely to be built?

A. point E

B. point A

C. point B

D. point D

____ 27. The steepest river gradient is most likely to be found at

A. point A

B. point D

C. point X

D. point Y

____ 28. In order to improve water navigation, the people in city X will most likely

A. construct levees

B. dredge river-bed sediments

C. restore wetlands along the river

D. construct locks and dams at point D

____ 29. The areas bounded by dashed lines represent regions. On what common characteristic are the regions most likely based?

A. elevation

B. precipitation

C. watershed

D. agriculture

____ 30. What is the approximate size of the Blue River drainage basin?

A. 6,000 square miles (15,540 sq. km.)

B. 17,000 square miles (44,000 sq. km.)

C. 28,000 square miles (72,500 sq. km.)

D. 36,000 square miles (93,000 sq. km.)

Telecommunications

Introduction

In this Challenge Project, you will use the information you gathered during your study of river and stream geography to analyze and evaluate this watershed (or another one) more fully. You will utilize a telecommunications system to collect additional data about the watershed conducted by other students at other places and times along rivers and streams, or to provide your observations and analysis to other students. You will select a particular topic relating to the available data. You will prepare visuals relevant to your topic, such as maps, tables, and graphs of your data. You may collect additional information from the library. Finally, you will interpret all the information that you have gathered and you will organize and write a paper about your findings. You may also work in groups to develop a map presentation to share the data and information you collect. You may also share some of your observations and research on your local river or stream with other students electronically.

Materials

Per group

- computer
- telecommunications system
- data from a rivers bulletin board or web site
- maps of rivers, streams, and watersheds being studied
- graph paper

Optional

- spreadsheet program for sorting and managing data

Procedure

PART A. Using Information from the Bulletin Board

1. Based on your geographic knowledge of a watershed, select as a group a river, stream, or other geographic area as the focus of your study. This may be your field-study site, local river or stream, another portion of the watershed, or even a different watershed that you want to compare or contrast with your own. Access the database to make sure the water system chosen has data available.
2. Based on your knowledge of available data, select a specific topic to investigate via the bulletin board. You may want to designate conditions to consider in your analysis, such as season of the year, weather conditions, or temperature.
3. Retrieve data for your selected area from the rivers bulletin board or web site. Check the amount of data available. If too little or too much information is available, modify the amount by changing the size of the system you are studying or limiting the time period for which you are collecting data.
4. Locate the test sites on maps and find out how the land is being used near each site. Note what communities, industrial areas, farming areas, or other land uses are nearby. To find this information, you may need to do research at a local library, look at other topographic maps, or refer to an atlas.
5. Devise visuals to support your work. If appropriate, devise a

scheme for mapping or graphing your data, comparing results to site location and land use.

6. Analyze your results, and prepare a presentation of your findings. In your presentation, take into account geographic concepts. Depending on the topic you select, questions to consider may include:
   - Describe the location of the site or area you have selected. What is its significant relative location?
   - What are the characteristics of this place?
   - What kinds of physical and human movement through this place have affected it?
   - What physical features affect the site or watershed you have selected?
   - What human factors affect this site or watershed?
   - Of what regions is this area a part? In what ways is this area representative of its region?
   - How does this area compare with your field-study site? Suggest reasons for the similarities and differences.
   - What might be done to improve or maintain environmental quality for this site?

PART B. Providing Information to the Bulletin Board

7. If instructed by your teacher, select a set of your observations or report you have created during *Rivers Geography* that would be useful to other students studying the watershed of which your river or stream is a part.
8. Prepare your material for transmission. This may include inputting it into your school's computer. Be sure to include basic facts about your information so it will be easy to use. As keywords, include the name of the river or stream, plus the watershed name. Also include absolute location of site, relative location of site, and date of observations. Give your information a clear title, so others will be able to quickly determine if they want this information for their purposes.
9. Follow your teacher's instructions for uplinking your data to the rivers bulletin board or other site, including naming your file.
10. To check the accessibility of your information, locate and download your file. Print out a copy and turn it in to your teacher.

Performance Criteria

A written report should:

- demonstrate that you can access (or upload) data electronically.
- show that you understand how to apply geographic concepts to real-world issues.
- show that you have researched and performed critical thinking in evaluating the relationship between geography and environment (or human behavior).
- include suggestions for improving or maintaining the water environment.
- reference resources used in preparing the report.

A group map presentation should:

- represent the efforts of everyone in the group.
- have one or more graphic displays of data (maps, charts, graphs) with accompanying explanations, sufficiently large to be clearly visible to the audience.
- include an oral presentation in which everyone in the group participates.
- be well-organized and use the map as a strong visual aid.
- include an opportunity for the audience to ask questions; all members of the group should respond to questions, especially those about their individual areas.

APPENDIX

B

Answers

Lesson 1 Answers to Questions

Student Information 1.1
Questions

1. Water on Earth exists as a solid, a liquid, and a gas (vapor). Liquid is the most common state.
2. The hydrologic cycle is the endless interchange of water among the oceans, the lower atmosphere, the land surface, and reserves several kilometers below Earth's surface.
3. Most of Earth's water (97 percent) is found in the oceans.
4. Precipitation and runoff fill rivers (surface streams), which then carry the water back to the oceans.
5. Answers will vary, but may include domestic (household), industrial, and municipal uses. Humans also use rivers for transportation, irrigation, and recreation.
6. Answers will vary but may include municipal sewage disposal, industrial water pollution, agricultural runoff, trash dumping, acid rain (caused by industrial and vehicle emissions), and residential pesticides and herbicides.

Student Activity 1.4
Observations

1. The bibliographical reference should be complete and in the form specified.
2. The problem or controversy will vary according to the article selected. Answers to parts a, b, and c should indicate whether a decision has or will be made, who the decision makers are, and alternative choices for decisions.

Analyses and Conclusions

1. The suggested title should be descriptive of the article's content.
2. Answers should identify whether the article is a news story or an editorial. The objectivity of the author can be evaluated by identifying the amount of space devoted to each point of view, the stated arguments for each point of view, and the prejudicial language used.
3. The author of the article may or may not formally cite sources. Quotations and statements from interviews should identify the authority used ("According to..."). The author may have conducted an on-site visit. Students may have to infer sources.
4. Answers will vary, but may include construction of locks and dams to facilitate shipping; dams to create reservoirs for recreation, control flooding, and ensure a constant water supply; dredging to improve navigation; draining wetlands for agriculture or building purposes; overdraft; pollution; construction of levees and floodwalls; and stabilization of banks. Effects of changes may be positive, negative, or both.
5. Answers will vary, but should include references to location, the human society, and an element of the natural river environment. A dam may have been constructed at a location where a substantial reservoir could be created; a lock and dam may have been built to provide a deeper channel for navigation; wetlands may have been drained and a levee constructed so that land could be used for a housing development, and so on.

   6a. Students should identify a past change that is contributing to a current problem. The change may involve the past modification of the environment for human use, or it could involve a condition that developed over time, such as an increasing population that necessitated construction of levees to protect property or drainage of wetlands for housing.

   b. Students should identify a change in the natural environment for the purpose of producing income or protecting a property that produces income.

7. In their summaries, students should identify the problem or controversy, the parties involved,

and the decisions to be made. The summary should include spatial, historic, and economic perspectives.

Critical Thinking Questions

1-3. Answers will vary.

Student Assessment 1.5
Observations

Student observations will vary with student interest in the topic and the medium used.

Analyses and Conclusions

Answers will vary with observation of the finished products and responses from other students.

Critical Thinking Questions

1. To some, the collage may have been a useful medium to express an idea or a feeling. Other students may be frustrated by their limited ability to present a written or verbal message.
2. Focusing attention on the river may help to clarify students' opinions and stimulate their curiosity. Lists will vary.

Lesson 2: Answers to Questions

Student Information 2.1
Questions

1. A river is a body of water that flows into a larger body of water. A river system consists of a river and tributaries that carry the water of a drainage basin.
2. Precipitation, temperature, and evaporation all influence the character of a river. Where precipitation exceeds evaporation, water is added to a river system. Where evaporation exceeds precipitation, water is removed from a river system. High temperatures increase evaporation.
3. The Amazon River has the greatest volume and discharge of all rivers because of its enormous watershed and heavy precipitation.
4. The Aswan High Dam controls the flooding of the Nile, produces water for irrigation, and converts moving water into electricity. It has created Lake Nasser, which is the world's largest reservoir, and it has reduced the current and volume of the Nile.
5. Bangladesh is frequently flooded because it is in area of heavy rainfall and it lies in the delta of the Ganges and Bramaputra Rivers.

Student Information 2.2
Questions

1. Compare student maps with Teacher Figure 2-1, at the end of Teacher Notes for Lesson 2. Each place name in the relative location descriptions (1–7) should be plotted and labeled. Precision should not be expected; however, sufficient clues are provided that students can demonstrate an understanding of relative location.
2. A map scale of 1:3,000,000 means that one map unit represents 3,000,000 of the same units on Earth's surface.
3. Cardinal directions refer to the four main directions: N, S, E, and W. Intermediate directions are more precise directions that fall between the four cardinal directions. Example of intermediate directions include NNE, NE, ENE, ESE, SE, SSE, SSW, SW, WSW, WNW, NW, and NNW.
4. The absolute location of Ābādān is 30.20 N, 48.16 E. Descriptions of relative location should include the following information: Ābādān is located in SW Iran on the Shatt al Arab River, which forms part of the Iran-Iraq border; it is approximately 50 km from the Persian Gulf in a petroleum-producing region in the Iranian province of Khuzistan on the delta formed by the Shatt al Arab and Kārūn Rivers.

Student Activity 2.3

Answers are given in corresponding Teacher Guide.

Student Assessment 2.4
Observations

Ho Chi Minh City (Saigon)	Station A
New Orleans	Station B
Vancouver	Station C
Dawson	Station D
Shanghai	Station E
Belém	Station F
Cairo	Station G
Buenos Aires	Station H

Analyses and Conclusions

1. Close inspection of temperature or precipitation should reveal correct matches, but several similar graphs may pose challenges to students.

   Ho Chi Minh City (formerly Saigon) and Belém have similar temperature ranges and total precipitation amounts, but several clues may be seen.

The slight temperature variation in Ho Chi Minh City can be attributed to its slightly higher latitude and the configuration of land masses.

Although located in a region that is constantly wet, Belém receives maximum precipitation from November to April. Ho Chi Minh City is more strongly influenced by seasonal wet and dry monsoons, and its maximum precipitation occurs from May to October. The Amazon has the largest volume.

Dawson and Cairo are both located in dry areas, but Dawson's much lower temperature graph should be obvious. Students may be surprised that the Yukon River has a greater discharge than the Nile.

The remaining four stations are all located in the middle latitudes and show similar temperature ranges. Some clues follow:

Vancouver and Buenos Aires both show precipitation maxima during the November–April period, but Buenos Aires' temperature graph indicates that it is located in the southern hemisphere. Furthermore, Río Paraná's discharge is much greater than the Fraser's.

Both the precipitation and temperature maps reveal that Shanghai has greater seasonal variability than New Orleans.

Lesson 3: Answers to Questions

Student Information 3.1
Questions

1. (a) Physical place characteristics should include bluff, oxbow lake, cutoff, floodplain, tributary, channel, bar, wetlands, lake, pond, marsh.
   (b) Human place characteristics should include riprap, reservoir, boat ramp, lock, dam, wing dam, and artificial levee.
2. Rivers are shaped by erosion and deposition and are agents of erosion and deposition.
3. Physical place characteristics that are the result of the abrupt slowing of a sediment-filled river are bars (or point bars), alluvial fans, and deltas.
4. Answers will vary but might include: dams, to provide hydroelectricity; and bridges to connect human activities on both sides of the river and facilitate transport of materials across the river.

Student Activity 3.2

Students should find 38 physical terms in the vocabulary list. Students should show 34 of these terms in their drawings and provide brief explanations of erosion, deposition, human activity, and gradient.

Student Information 3.3
Questions

1. Topographic maps show physical and human features that appear on the earth's surface. Physical features include water (such as streams, lakes, ponds, reservoirs, swamps, and marshes), configuration of the land surface (shown by contour lines and spot elevations), and forests. Human features include buildings, transportation and communication lines; human activities, such as mining; and alterations of physical features, such as dams, levees, bridges, road cuts, and fill.
2. A 7.5-minute series map (scale 1: 24,000) would be appropriate for a detailed analysis of an area. Each map of this series shows 7.5 minutes of latitude and longitude.
3. Point, line, and area symbols are used. These symbols may represent actual physical objects that are visible in the landscape or they may symbolize features that exist only through generally accepted human knowledge.
4. Contour lines on USGS maps are brown lines that represent the height of land surfaces relative to sea level. To determine a river's gradient, calculate the difference in elevation between two lines that cross the river; then divide that difference by the stream distance between the two lines.

Student Activity 3.4
Analysis and Conclusions

1. The map shows an island in a lake. The island has a generally elongated spherical form, longest from N to S. The coast line is regular or smooth. A single permanent stream originates in the center of the island and flows to the S coast. A hill rises to a height of 507 feet in the east central part of the island. With the exception of the E slope of the hill, which assumes a sharp cliff or bluff, most slopes are relatively gentle. See sample contour lines and elevation map on next page.

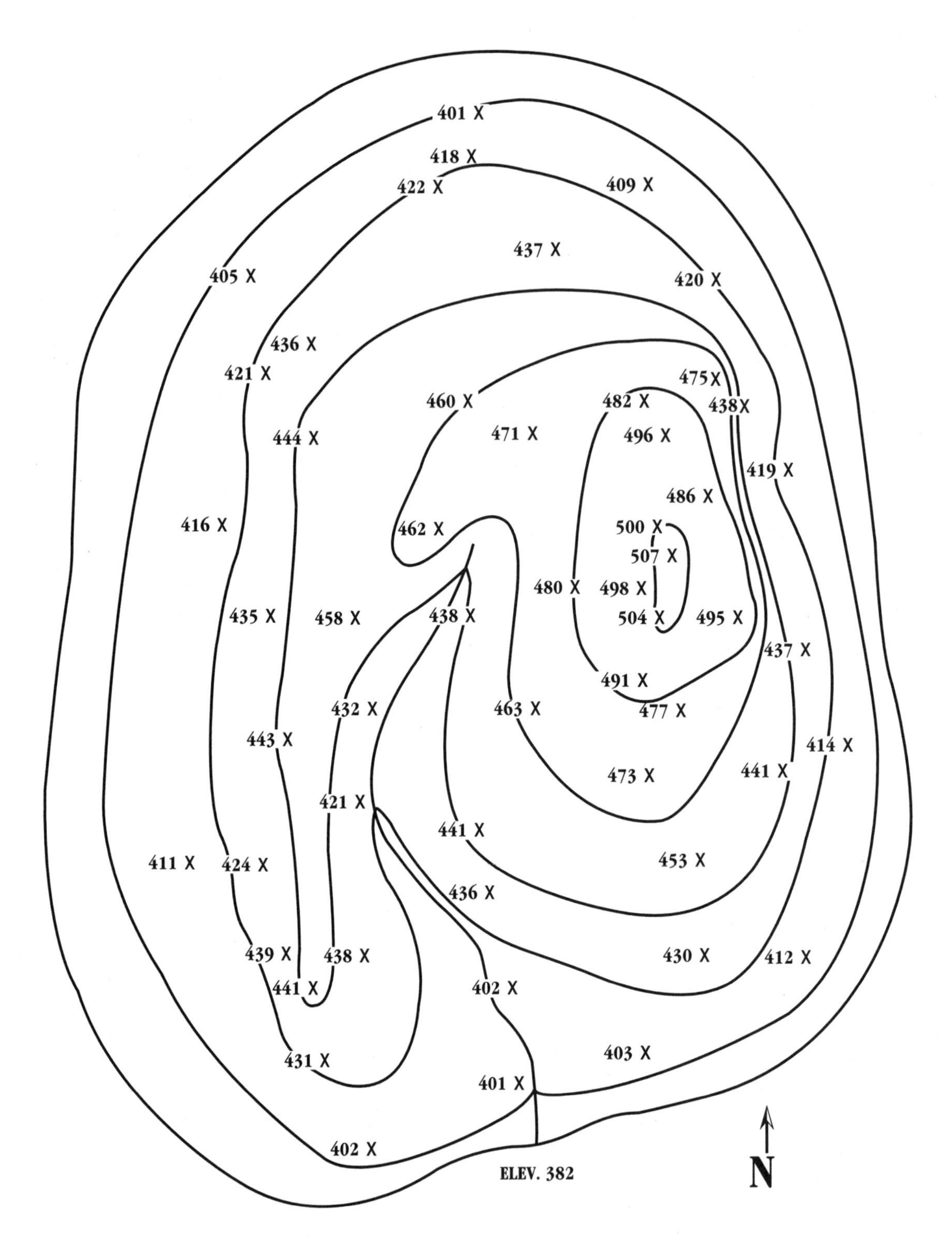

Appendix Figure B-1: Island Map With Point Elevations and Contour Lines

2. The local relief is 125 feet from the top of the hill to the coastline (507 feet – 382 feet = 125 feet). In metric measure, the local relief is 38 meters (125 × 0.30 = 38 meters).

Student Activity 3.5

See corresponding Teacher Guide for sample observations and answers.

Lesson 4: Answers to Questions

Student Information 4.1
Questions

1. Ideas about how people should interact with one another and with the environment are expressed as material culture, including all that is visible in the human landscape. Architecture, clothing, street patterns, customs, and celebrations will likely be cited.
2. Answers will depend on the community. Students may recognize cultural practices that differ from those in other communities, or they may identify ethnic neighborhoods and ethnic groups. They may also observe differences in customs and celebrations between longtime residents and newcomers.
3. A family tree is a genealogical chart illustrating the relationships of family members to one another and to their ancestors. By plotting the birthplaces of class members and their families on a map, observations can be made about spatial distributions and diffusions.
4. Genealogical information for constructing a family tree can be obtained from family members, family records and written histories, and unofficial family historians. More detailed research can be conducted by investigating public records and information in genealogical libraries.
5. Sources of information may include school and community libraries, teachers, family members, neighbors, historical societies, city government officials, and city government personnel.

Student Activity 4.2

See corresponding Teacher Guide for sample observations and answers.

Student Information 4.3
Questions

1. Any form of communication and transportation involves movement. People travel from homes to jobs, school, or to engage in recreation. Goods are moved from place to place to satisfy people's needs and wants, and ideas are exchanged through public and private channels, including conversations. All show interrelationships and interdependence.
2. Nonmaterial culture refers to intangible elements, such as ideas, customs, traditions, and beliefs. Material culture involves tangible objects, but is closely tied to nonmaterial culture because the objects are designed to express ideas or to accomplish objectives important to the culture. Cultural landscape is the distinctive appearance of a particular area, a product of the culture practiced there.
3. Rivers have historically served as transportation routes for explorers and later as routes taken by immigrants. Sometimes early settlers found rivers barriers to movement. Today, many rivers still provide transportation routes for the movement of goods and people.
4. People emigrate because of "push" factors—the desire to escape undesirable conditions—or "pull" factors—the desire to take advantage of perceived or real opportunities elsewhere. Many times, emigration occurs because of a combination of such factors.
5. In a "melting pot," a new culture develops from a blending of elements from two or more cultures. In other cases, cultures may remain distinct from one another, leading to a "mosaic" of several or many different cultures.

Lesson 5: Answers to Questions

Student Information 5.1
Questions

1. Humans decide to modify the physical environment when they perceive that the changes will improve living conditions. Cultural values, available technology, and funding influence such decisions.

2. Although rivers serve a natural system, they may not serve human needs efficiently. To meet human needs, flooding must be controlled so that the floodplain can be used for human activities, and channels must be dredged and straightened to provide for transportation. Using river water for agricultural irrigation (and other uses) often requires transporting it to areas far from the riverbanks.
3. Humans build structures to control flooding and erosion, provide hydroelectricity, and improve navigation. They also create reservoirs, pipelines, and canals for a consistent water supply. Humans also dredge rivers to maintain navigation channels, and they may use rivers as waste depositories. Activities that indirectly affect rivers include paving land surfaces, removing vegetation, and draining wetlands.
4. Humans may fail to consider the long-range impact of changes. In particular, humans may not fully consider the ecological perspective when planning changes intended to control the river or harness its potential to provide economic benefits.

Student Information 5.2
Questions

1a. Special-interest group: people bound together to express their collective preference for particular interests.
b. Conservation: measures that preserve or restore the natural environment.
c. NIMBY: acronym for "Not In My Backyard," an attitude held by people who oppose the development of some project in the vicinity of their homes.
d. Waste: materials discarded because they are perceived to be of no value.
e. Recycling: reuse of materials commonly classified as waste; the conversion of waste materials to useful products.
f. Sanitary landfill: a specially prepared site used as a depository for waste materials.
g. Region: an area differentiated from other areas according to specified criteria.
h. Environmental impact study: an investigation that assesses the impact of human modifications of the physical and human environments.

2. Answers will vary but may include any of the following:
   - One of the human modifications of the river is dumping waste into rivers. It is an inexpensive, convenient way of disposing of wastes; but it also pollutes water, destroys aquatic life, is aesthetically displeasing, and limits use of water for domestic and recreational purposes.
   - One of the human modifications of the river is building dams for hydroelectricity. It provides relatively inexpensive, nonpolluting source of power; but it also alters the natural system by impounding much of a river's water in a reservoir, increases sedimentation, and disrupts ecosystems.
   - One of the human modifications of the river is channelization for transportation. It creates a shorter, more efficient route, but it also may destroy wetlands and hasten the river's flow by shortening its channel.
   - One of the human modifications of the river is building levees. It protects floodplains for human use; but it also eliminates the floodplain as a reservoir for the river during times of flood, destroys wetlands on floodplain, and causes higher river stages by restricting river to channel only.
   - One of the human modifications of the river is draining of wetlands. It provides additional land for agriculture, and for construction of houses, businesses, and transportation routes; but it also destroys wetland ecological systems and hastens runoff into rivers, causing higher flood stages.
3. People disagree about how to use rivers because of differences in cultural backgrounds, economic interests, knowledge of and interest in natural systems, and proximity to rivers. Examples will vary and may include any of the following: A person who lives on the floodplain will want a levee to protect his or her home. A barge operator will favor dredging the river to maintain a navigation channel. A boat owner may support the development of an oxbow lake into a marina. A city government may favor plans to stabilize its waterfront with riprap or concrete embankments.

An engineering firm may want to straighten tributaries and fill in floodplain wetlands to construct a highway.

4. Answers will vary, but may include any of the following pairs of opponents: (a) A naturalist wants to preserve wetlands on a floodplain, while a housing developer wants to drain the wetlands. (b) One biologist wants to introduce new varieties of fish to promote recreational fishing, while another opposes the plan because the ecology will become imbalanced. (c) One person wants to prevent human visitors in a natural area, while another wants to permit boaters to observe and appreciate the area's ecology and beauty. (d) Administrators of a reservoir want to keep the reservoir full to provide a constant water supply for communities and to prevent disruption of the lake's ecology and recreational activities, while officials downstream are concerned that their water supply will suffer from diminished stream flow.

Student Activity 5.3

See corresponding Teacher Guide for sample observations and answers.

Lesson 6: Answers to Questions

Student Information 6.2
Questions

1. (a) Location describes where some phenomenon is. (b) Place describes physical and human features. (c) Movement investigates spatial interaction. (d) Human-environment interaction investigates human modifications of nature. (e) Region is an area that is distinct from other areas.
2. Location, place, movement, and human-environmental interaction have been studied so far. Lesson 7 focuses on region.
3. A field study features the real world rather than symbolic representations of the world. While the first five lessons were analytical, the field studies in this lesson will be partly analytical and partly synthetic.
4. Analysis requires that you study a phenomenon by separating it into its component parts. Synthesis requires that you combine all the parts to show a larger picture.
5. Answers will vary.

Lesson 7: Answers to Questions

Student Information 7.1
Questions

1. Answers will vary in form, but they should refer to the two distinguishing features of a region: It occupies a particular area, and it is in some way distinct from other regions. Many examples are possible, included those noted in the information sheet.
2. Both a uniform region and a functional region occupy areas and both are distinguishable from other regions. A uniform region possesses some characteristic that is found more or less uniformly throughout. A functional region has a focal point that is the center of some regionwide activity, such as a school, grocery story, or factory.

Student Assessment 7.2
Analyses and Conclusions

1. Selection of a topic will vary with student interest. Popular topics include fishing, boating, and pollution. Some may be influenced by the field-study experience.
2. Selection of region will vary; definition should be based on specific characteristics. Students who choose the field-study site as their region must still define that region specifically.
3. Students should have used the five themes as a framework. Other concepts may include an understanding of analysis and synthesis, human modifications of the physical landscape, conflicting interests regarding land use, and decisionmaking about the environment. The hydrologic cycle, the ecological importance of wetlands, and the effects of pollution may also be mentioned. The selection of concepts will be influenced by the topic chosen by each student.
4. Answer will depend on student's topic and regional definition.

Critical Thinking Questions

1. Both the brochure and the collage can be useful in presenting an idea. More specific information can be presented in a brochure, whereas a collage is more likely to create an impression.
2. Students should be able to identify general groups, such as agriculture, business, conservation, or recreation. Identification of groups who agree or disagree with each student's brochure will aid in values clarification.
3. Studying a topic in a particular region is more useful than studying that topic worldwide because a geographer can study a particular region in greater detail. This allows not only consideration of a larger number of topics but also attainment of a better understanding of their relationships. Once a particular region has been studied, it can be compared with other similar regions.

APPENDIX C

Resources

Organizations

In addition to your state, local, and regional offices of geological survey, energy, natural resources, and, soil and water conservation, the following organizations and federal government agencies can provide useful information for use in the classroom.

AAA Publishing, 1000 AAA Drive, M/S 65, Heathrow, FL 32746-5063. Phone: 407-444-8201. American Automobile Association's "How to Read a Map" Program Kit.

Academy of Natural Sciences, 1900 Benjamin Franklin Parkway, Philadelphia, PA 19103-1195.

Adopt-A-Stream, P.O. Box 435, Pittsford, NY 14534-0435. Water-quality monitoring, clean-up programs.

American Forests, 1516 P Street NW, Washington, DC 20005. Phone: 202-667-3300.

American Rivers, 801 Pennsylvania Avenue SE, Suite 400G, Washington, DC 20003-2167. Preservation and restoration of rivers.

American Water Resources Association, 5410 Grosvenor Lane, Suite 220, Bethesda, MD 20814-2192. Posters and booklets.

American Water Works Association, 6666 West Quincy Avenue, Denver, CO 80235-3098. Campaign to preserve water resources.

America's Clean Water Foundation, 750 First Street NE, Suite 911, Washington, DC 20002-4241. Educational materials.

Center for Environmental Information, 46 Prince Street, Rochester, NY 14607. Phone: 716-271-3550. Great Lakes information.

Center for the Great Lakes, 35 E. Wacker Drive, Suite 1870, Chicago, IL 60601.

Center for Marine Conservation, 1725 DeSales Street NW, Suite 500, Washington, DC 20036. Phone: 202-429-5609.

Earth Science Information Center, U.S. Geological Survey, 507 National Center, Reston, VA 22092. Phone: 703-648-6892 or 800-USA-MAPS. Request list of publications. Classroom map lesson packets include *Exploring Maps* and *What Do Maps Show?*

Environment Canada, Great Lakes Environment Program, 25 St. Clair Avenue E., Toronto, ON M4T 1M2. Phone: 416-973-6467.

Environmental Protection Agency, 401 M Street SW, Washington, DC 20460.

EPA Wetlands Information Hotline: 800-832-7828. Mon.–Fri. 9:00 A.M.–5:30 P.M. EST.

Freshwater Foundation, 725 County Road 6, Wayzata, MN 55391-9611. Freshwater research.

Geography Education National Implementation Project (GENIP), 1710 Sixteenth Street NW, Washington, DC 20009. Phone: 202-234-1450. Publications are distributed through National Council for Geographic Education.

Great Lakes Historical Society, P.O. Box 435, 480 Main Street, Vermilion, OH 44089. Phone: 216-967-3467.

Greenpeace USA, 1436 U Street NW, Washington, DC 20009. Phone: 202-462-1177.

International Joint Commission, 100 Ouellette Avenue, 8th Floor, Windsor, ON N9A 6T3. Phone: 519-257-6700.

International Joint Commission, Great Lakes Regional Office, P.O. Box 32869, Detroit, MI 48232. Phone: 313-226-2170.

Izaak Walton League of America, 1401 Wilson Boulevard, Level B, Arlington, VA 22209-2318. "Save Our Streams" program, publications.

Lake Michigan Federation, 59 E. Van Buren Street, Suite 2215, Chicago, IL 60605. Phone: 312-939-0838.

National Arbor Day Foundation, The, 100 Arbor Avenue, Nebraska City, NE 68410. Phone: 402-474-5655.

National Audubon Society, National Education Office, Rt. 4, Box 171, Sharon, CT 06069.

National Council for Geographic Education, 16A Leonard Hall, Indiana University of Pennsylvania, Indiana, PA 15705-1087. Phone: 412-357-6290. Publishes *Key to the National Geography Standards, Outline Map Blackline Masters* (set of 15 base maps), and bimonthly *Journal of Geography* and *Perspective* newsletter. Send for lists of publications, map product sources, software, and games and activities. Ask for Environmental Information list.

National Geographic Society, 1145 17th Street NW, Washington, DC 20036. Web site: http://www.nationalgeographic.com. Publishes *National Geographic, National Geographic World,* and *National Geographic Traveler.*

———. Educational Services, P.O. Box 98019, Washington, DC 20090-8019. Phone: 800-368-2728. Classroom materials: printed materials, maps, atlases, films, filmstrips, telecommunications, videodisks; poster, *Maps, The Landscape, and Fundamental Themes of Geography.*

———. Geography Education Program: Geographic Alliances, P.O. Box 98190, Washington, D.C., 20090-8190. Phone: 202-775-6701. Provides matching funds to promote geography education through state alliances, via such programs as summer programs for teachers, Geography Awareness Week, and the National Geography Bee. Contact for name of your state's alliance coordinator, who can put you on mailing list for state alliance newsletter, workshop announcements, and other descriptions of teaching strategies and classroom materials.

National Marine Fisheries Service, U.S. Department of Commerce, Washington, DC 20036.

National Wildlife Federation, Great Lakes Natural Resource Center, 506 E. Liberty Street, 2nd Floor, Ann Arbor, MI 48104-2210. Phone: 313-769-3351.

Natural Resources Defense Council, National Office, 1350 New York Avenue NW, Suite 300, Washington, DC 20005. Phone: 202-783-7800.

Natural Resources Defense Council, Public Information, 40 West 20th Street, New York, NY 10011. Phone: 212-727-2700.

Nature Conservancy, The. 1815 North Lynn Street, Arlington, VA 22209. Phone: 703-841-5300.

Population Reference Bureau, Inc., 1875 Connecticut Avenue NW, Suite 520, Washington, DC 20009. Phone: 202-483-1100. Print materials and film catalog.

Sierra Club, Information Center, 730 Polk Street, San Francisco, CA, 94109, Phone: 415-776-2211. See *Green Guide* in list of print resources.

U.S. Army Corps of Engineers, Regulatory Branch, CECW-OR, 20 Massachusetts Avenue NW, Washington, DC 20314-1000.

U.S. Environmental Protection Agency, Great Lakes National Program Office, 77 W. Jackson Boulevard, Chicago, IL 60604. Phone: 800-621-8431 from IN, MI, MN, OH, WI, IL; 312-353-2072 from Chicago and other parts of U.S.

U.S. Fish and Wildlife Service, Department of the Interior, Washington, DC 20240. Information on wetlands.

U.S. Geological Survey. National Water Information Clearinghouse, USGS, 423 National Center, Reston, VA 22092-0002. Phone: 800-426-9000

———. National Water Quality Assessment Program: Deputy Assistant Chief Hydrologist NAWQA Program, USGS, National Center, 12201 Sunrise Valley Drive, MS 413, Reston, VA 22092.

———. Public Inquiries Office, 503 National Center, Room 1-C-402, 12201 Sunrise Valley Drive, Reston, VA 22092. Phone: 703-648-6892; 800-USA-MAPS.

———. USGS Map Sales, Federal Center, Box 25286, Denver, CO 80225. Phone: 303-236-7477. Map-ordering information.

University Corporation for Atmospheric Research, Office for Interdisciplinary Earth Studies, P.O. Box 3000, Boulder, CO 80307-3000.

Water Environment Federation, 601 Wythe Street, Alexandria, VA 22314-1994. Materials on water-quality issues.

Wilderness Society, The, 1900 17th Street, Washington, DC 20006. Phone: 202-833-2300.

World Wildlife Fund, 1250 24th Street NW, Suite 500, Washington, DC 20037. Phone: 202-293-4800.

Zero Population Growth, 1400 16th Street NW, Suite 320, Washington, DC 20036. Phone: 202-332-2200.

Books and Classroom Kits

American Automobile Association, "How to Read a Map" Program kit, Heathrow, FL: AAA Publishing.

Anderson, Jeremy. *Teaching Map Skills: An Inductive Approach.* Indiana, PA: National Council for Geographic Education, 1986.

Benhart, John E. and Alex R. Margin, Jr. *Wetlands: Science, Politics, and Geographical Relationships.* Indiana, PA: National Council for Geographic Education, 1994.

Brown, Lester R., Christopher Flavin and Hal Kane. *Vital Signs 1996: The Trends That Are Shaping Our Future.* New York: W. W. Norton, 1996.

Christopherson, Robert W. *Geosystems: An Introduction to Physical Geography.* New York: Macmillan, 1992.

Costner, Pat and Joe Thornton. *We All Live Downstream: The Mississippi River and the National Toxics Crisis.* Washington, DC: Greenpeace, 1989.

Crews, Kimberly A. and Patricia Cancellier. *Connections: Linking Population With the Environment.* Washington: The Population Reference Bureau, Inc., 1991. Teaching kit: Teacher's Guide, Student Resource Book.

Espenshade, Edward B., ed., *Goode's World Atlas,* 19th Edition. New York: Rand McNally and Company, 1995.

Fromboluti, Carol Sue. *Helping Your Child Learn Geography.* U.S. Department of Education. Order from: Geography, Consumer Information Center, Pueblo, CO 81009.

Garrett, Wilbur E. (ed.) *Historical Atlas of the United States.* Washington, DC: National Geographic Society, 1988.

Geography Educational National Implementation Project. *K–6 Geography: Themes, Key Ideas and Learning Opportunities.* Indiana, PA: National Council for Geographic Education (NCGE), 1987.

———. *7–12 Geography: Themes, Key Ideas, and Learning Opportunities.* Indiana, PA: NCGE, 1989.

Geography Education Standards Project. *Geography for Life: National Geography Standards 1994.* Washington, DC: National Geographic Research & Exploration, 1994, pdt #01775-1216. Send $9 to National Geography Standards, P.O. Box 1640, Washington, DC 20013-1640.

Gersmehl, Phil. *The Language of Maps.* Pathways in Geography Series #1. Indiana, PA: National Council for Geographic Education, 1991.

Izaak Walton League. *Save Our Streams.* Arlington, VA: Izaak Walton League of America.

Kidron, Michael and Ronald Segal. *The State of the World Atlas.* 5th ed., Myriad Editions, Ltd., 1995.

Ludwig, Gail S., et al. *Directions in Geography: A Guide for Teachers.* Washington, DC: National Geographic Society, 1991. K–12 resource; lesson plans focusing on 5 themes, blackline maps, annotated bibliography.

Mitchell, Mark K. and William B. Stapp. *Field Manual for Water Quality Monitoring*, 9th ed. Dexter, MI: Thomson-Shore, 1995. Order from: William B. Stapp, 2050 Delaware Ave., Ann Arbor, MI 48103.

Nace, Raymond L., *Water of the World*. USGS booklet 1984-421-618/017, USGS, Water Resources Division.

National Council for Geographic Education (NCGE). *Guidelines for Geographic Education: Elementary and Secondary Schools*. Joint Committee of NCGE and Association of American Cartographers.

———. *Key to the National Geography Standards*. 1995.

———. *Outline Map Blackline Masters*. Set of 15 base maps. ($5 plus $2 shipping and handling.)

———. *Spaces and Places: A Manual for Geography Teachers*. K–12 guide for implementing geography standards.

National Geographic Atlas of the World. 6th ed. Washington, DC: National Geographic Society, 1990.

National Geographic Atlas of North America: Space Age Portrait of a Continent, The. Washington, DC: National Geographic Society, 1985.

New York Times Atlas of the World, The. Third revised concise edition. New York: Times Books, 1992.

Nystrom Desk Atlas. Chicago: Nystrom, 1994.

Pitzl, Gerald, ed. *Geography 96/97*, et al. Guilford, CT: Dushkin. Annual publication of reprints of significant geographic articles.

Posey-Pacak, Melissa L. *Earth at Risk: Instructional Materials on the Sustainable Development and Management of the Environment*. Indiana, PA: National Council for Geographic Education. Annotated resource guide.

Rand McNally. *Atlas of Oceans*. Chicago: Rand McNally and Company, 1993.

Reighton, James F. *Public Involvement Manual: Involving the Public in Water and Power Resources Decisions*. U.S. Government Printing Office. Washington, DC: United States Department of the Interior, 1980.

Robinson, Ann and Robbin Marks. *Restoring the Big River: A Clean Water Act Blueprint for the Mississippi*. Izaak Walton League of America and Natural Resources Defense Council, February 1994. Contact Izaak Walton League of America, 5701 Normandale Road, Suite 5701, Minneapolis, MN 55424, Phone: 612-922-1608 or National Office, Natural Resources Defense Council.

Salter, Christopher L., Gail L. Hobbs, and Cathy Salter. *Key to the National Geography Standards*. Washington, DC: National Geographic Society. 1994.; available through NCGE.

Sierra Club. *Green Guide*. San Francisco: Sierra Club. Education guide to free and inexpensive environmental materials.

Slater, Frances. *Learning Through Geography*. Pathways in Geography Series. Indiana, PA: National Council for Geographic Education.

Social Studies Resources Series, Inc. Reprinted periodical articles. The following volumes relate to *Rivers Geography: Energy, Habitat, Pollution, Population, Technology, Third World, Transportation, Earth Science, Life Science,* and *Physical Science.*

Spaulding, Nancy E. *Earth Science Laboratory Investigations: Teacher's Annotated Edition.* Lexington, MA: D.C. Heath & Co., 1989.

Strahler, Arthur N. and Alan H. Strahler. *Elements of Physical Geography.* 4th ed. New York: John Wiley and Sons, 1989.

Terrell, Charles R. and Patricia Bytnar Perfetti. *Water Quality Indicators Guide: Surface Waters.* U.S. Department of Agriculture Soil Conservation Service, September, 1989.

Times Atlas of the World, The. London: Times Books in collaboration with John Bartholomew & Son Limited, 1977.

U.S. Environmental Protection Agency. *Great Minds? Great Lakes!* EPA Document No. 905/M/90/004. Contact EPA office.

———. *America's Wetlands: Our Vital Link Between Land and Water.* EPA-87-016.

———. *National Directory of Citizen Volunteer Environmental Monitoring Programs.* EPA503/9-90-004.

———. *The Watershed Protection Approach: An Overview.* EPA/503/9-92/002.

U.S. Geological Survey. *Popular Publications of the U.S. Geological Survey.* Lists the leaflets *What Is Water?* and *The Hydrologic Cycle* as well as other publications. Write for catalog of additional titles in the series from Books and Open-file Reports Section, U.S. Geological Survey, Federal Center, Box 25425, Denver, CO 80225.

Van der Leeden, Frits, Fred Troise, and David K. Todd. *The Water Encyclopedia.* Chelsea, MI: Lewis Publishers, 1991.

Waterstone, Marvin. *Water in the Global Environment.* Pathways in Geography Series #3. Indiana, PA: National Council for Geographic Education.

Winston, Barbara J. *Map and Globe Skills: K–8 Teaching Guide.* Indiana, PA: National Council for Geographic Education, 1986.

World Book Atlas, The. Chicago: World Book, Inc., 1986.

World Resources Institute. *Information Please Environmental Almanac.* Boston: Houghton Mifflin Company.

Periodicals

EarthQuest. Office for Interdisciplinary Earth Studies, University Corporation for Atmospheric Research, P.O. Box 3000, Boulder, CO 80307-3000. Quarterly publication.

EPA Journal. Superintendent of Documents, Government Printing Office, Washington, DC 20402.

Focus. American Geographic Society, 156 Fifth Avenue, Suite 600, New York, NY 10010-7002. Phone: 212-242-0214. Quarterly.

Journal of Geography. National Council for Geographic Education, 16A Leonard Hall, Indiana University of Pennsylvania, Indiana, PA 15705-1087. Phone: 412-357-6290. Bimonthly.

National Geographic. National Geographic Society (NGS), 1145 17th Street NW, Washington, DC 20036. For recent and useful articles, consult an NGS index at a library.

National Geographic Traveler. National Geographic Society, 1145 17th Street NW, Washington, DC 20036.

National Geographic World. National Geographic Society, 1145 17th Street NW, Washington, DC 20036.

Perspective. National Council for Geographic Education, 16A Leonard Hall, Indiana University of Pennsylvania, Indiana, PA 15705-1087. Phone: 412-357-6290. Bimonthly newsletter.

Science. American Association for the Advancement of Science, 1333 H St., NW, Washington, DC 20005.

Scientific American. 415 Madison Avenue, NY 10017.

Worldwatch. Worldwatch Institute, 1776 Massachusetts Avenue NW, Washington, DC 20036. Phone: 202-452-1999.

Media

Agency for Instructional Technology (AIT). *Geography in U.S. History*. A series of ten 20-minute video programs. AIT, Box A, Bloomington, IN 47402-0120. Phone: 800-457-4509.

———. *Global Geography*. A series of ten 15-minute video programs focusing on geography's themes. Teacher's guide. Developed by Agency for Instructional Technology in association with 31 state and Canadian provincial agencies. Contact your state department of education or AIT.

National Council for Geographic Education. Slide sets on various topics.

National Geographic Society, Educational Services, P.O. Box 98019, Washington, DC 20090-8019. Phone: 800-368-2728. Maps, films, filmstrips, telecommunications, videodisks. Notable items include: "The Island" (14 min., concern for environment); "Geography: A Voyage of Discovery" (15 min., student introduction to geography); "Reflections on Water" (14 min., introduction to water resources and issues.)

Video Project, The. 5332 College Avenue, Suite 101, Oakland, CA 94618. Phone: 800-4-PLANET. Film: "The Mighty River," 24 minutes.

Appendix D

The National Geography Standards

Physical and human phenomena are spatially distributed over Earth's surface. The outcome of *Geography for Life* is a geographically informed person (1) who sees meaning in the arrangement of things in space; (2) who sees relations between people, places, and environments; (3) who uses geographic skills; and (4) who applies spatial and ecological perspectives to life situations.

The World in Spatial Terms

Geography studies the relationships between people, places, and environments by mapping information about them into a spatial context.

The geographically informed person knows and understands:

1. How to use maps and other geographic representations, tools, and technologies to acquire, process, and report information from a spatial perspective
2. How to use mental maps to organize information about people, places, and environments in a spatial context
3. How to analyze the spatial organization of people, places, and environments on Earth's surface

Places and Regions

The identities and lives of individuals and peoples are rooted in particular places and in those human constructs called regions.

The geographically informed person knows and understands:

4. The physical and human characteristics of places
5. That people create regions to interpret Earth's complexity
6. How culture and experience influence people's perceptions of places and regions

Physical Systems

Physical processes shape Earth's surface and interact with plant and animal life to create, sustain, and modify ecosystems.

The geographically informed person knows and understands:

7. The physical processes that shape the patterns of Earth's surface
8. The characteristics and spatial distribution of ecosystems on Earth's surface

Human Systems

People are central to geography in that human activities help shape Earth's surface, human settlements and structures are part of Earth's surface, and humans compete for control of Earth's surface.

The geographically informed person knows and understands:

9. The characteristics, distribution, and migration of human populations on Earth's surface
10. The characteristics, distribution, and complexity of Earth's cultural mosaics
11. The patterns and networks of economic interdependence on Earth's surface
12. The processes, patterns, and functions of human settlement
13. How the forces of cooperation and conflict among people influence the division and control of Earth's surface

Environment and Society

The physical environment is modified by human activities, largely as a consequence of the ways in which human societies value and use Earth's natural resources, and human activities are also influenced by Earth's physical features and processes.

The geographically informed person knows and understands:

14. How human actions modify the physical environment
15. How physical systems affect human systems
16. The changes that occur in the meaning, use, distribution, and importance of resources

The Uses of Geography

Knowledge of geography enables people to develop an understanding of the relationships between people, places, and environments over time—that is, of Earth as it was, is, and might be.

The geographically informed person knows and understands:

17. How to apply geography to interpret the past
18. How to apply geography to interpret the present and plan for the future

---

Excerpted from *Geography for Life: National Geography Standards 1994,* The Geography Education Standards Project. Washington, DC: National Geographic Research and Exploration, 1994, pp. 34–35. Used by permission.

GLOSSARY

absolute location a precise and unique position on earth, usually indicated by latitude and longitude.

alluvial fan a fan-shaped deposit of rock and sediment, deposited by a water flow whose velocity has been abruptly decreased because of a change in gradient, movement to a less constricted channel, or both.

alluvium (uh LOO vee uhm) silt, sand, and fine rock particles deposited by a river on floodplains and deltas.

analysis the process of separating a phenomenon into parts for the purpose of closer examination.

ancestor relative from whom an individual is descended and who is more remote than a grandparent.

architecture (AHR kuh tehk cher) design and style of buildings.

area symbol a map symbol that represents the space on Earth's surface occupied by a particular phenomenon.

artificial levee a raised bank along a river, built by humans to protect settlements, transportation, or economic activities from floods.

atmosphere the thin layer of gases enveloping earth; the atmosphere is densest at the earth's surface; 97 percent of the atmosphere lies within 16 km (10 miles) of the earth's surface.

bank the sloping ground along the margin of a river or stream marking the boundary of the channel.

bar an accumulation of sand and gravel within a river channel due to a change in river velocity, as a point bar on the inside curve of a meander.

bed the bottom of a stream or river channel.

bench mark a spot elevation shown on a topographic map by the abbreviation "BM," and an "X" to mark the spot; the number indicates the elevation above or below sea level.

bluff a cliff with a broad front bordering a river, formed by the action of the river cutting into the valley walls.

boat ramp a concrete or gravel surface extending from the river bank onto the river bed for the purpose of launching or removing boats.

braided river a network of channels flowing around many bars in the middle of the river.

bridge a structure spanning a river or other obstruction to allow movement across that feature.

cardinal directions north, south, east, and west. (Compare with intermediate directions.)

channel the bed of a stream; the course through which a stream flows.

climograph a graphic illustration of temperature and precipitation averages for a particular weather station.

commerce trade; buying, transporting, and selling goods and services.

community a relatively permanent cluster of people who have gathered to satisfy basic needs and wants.

compass direction the general direction of travel shown by a compass, such as NE, S, WSW; compass heading or bearing is more precise, for example, 92 degrees, 121 degrees, and so on.

condensation (kahn dehn SAY shuhn) the process by which water vapor changes to liquid, caused by cooling or air saturation.

conservation attempts to achieve a favorable balance between human and physical elements by preserving natural resources from destruction.

contour interval the vertical distance between adjacent contour lines.

contour lines or **contours** imaginary lines that connect points of equal elevation on topographic maps; isopleth lines are brown and connect points of equal value.

course the path followed by a river.

cultural diffusion the spread of cultural traits from an area where they are practiced to another area.

cultural feature element of the environment that has been constructed by humans and that has in some way modified the physical environment.

cultural landscape a landscape that has human features; the distinctive appearance a human group makes on the area it occupies.

cultural pluralism the existence of two or more distinct culture groups at the same time in the same region.

culture the unique set of qualities, such as language, customs, religion, and traditions that distinguishes one group of people from another.

culture area See cultural island.

culture island an area within a larger culture where a smaller but distinct culture is practiced.

cut bank that portion of a river bank most subjected to the erosive action of the river; it occurs on the outer edge of a meander curve, where velocity is greatest.

cutoff the new river channel that is formed when the river cuts through the neck of a meander.

dam an obstruction, either naturally occurring or constructed by humans, that impounds water on the upstream side. Human-constructed dams are for the purposes of controlling floods, producing electric power, aiding transportation, or storing water for irrigation, drinking, or recreation.

data (plural of datum) information in the form of facts or figures.

delta a deposit of soil (often sand) at the mouth of a river.

deposition (dehp uh ZISH uhn) the process by which eroded materials are laid down.

diffusion the "borrowing" or spread of a physical or human feature from one location or region to another.

discharge the amount of water flowing down a stream or river into a larger body of water, dependent upon velocity (speed) and volume (quantity) and measured in cubic meters per second or cubic feet per second.

distance decay the decrease in frequency of occurrence of a particular phenomenon as distance from a central point increases.

divide a ridge or relatively high ground that separates two drainage basins.

dock a platform on or above the river surface, extending from the bank to provide access to boats and a "tie-up" for boats.

drainage basin the area drained by a single river system; a watershed.

elevation the height or altitude above some particular level; on topographic maps, the height of land above or the depth below sea level.

emigration (em uh GRAY shuhn) the process of leaving one's native country to settle in another.

environmental impact study an assessment of the impacts of specific human modifications on the physical environment and on the human landscape.

environmental modification a change made by humans to earth's natural features.

equator the parallel of latitude that circles earth exactly midway between the North Pole and the South Pole; 0 degrees latitude.

erosion (ih ROH zhuhn) the process by which land surfaces are worn away by the actions of running water, wind, and ice.

ethnic group people who share a common culture and who may have distinct physical characteristics.

evaporation (ih vap uh RAY shuhn) the process by which liquid water changes to a gaseous state (vapor).

evapotranspiration (ih VAP oh tran spuh RAY shuhn) the change of water into vapor by evaporation from land and water surfaces and by transpiration from plants.

exotic river a river that flows through a desert.

family tree a sequential arrangement of the members of a family traced back from those now living to their ancestors.

ferry a vessel that transports passengers and vehicles across a river or other body of water.

field study the gathering and analysis of first-hand information by systematically applying concepts and skills to observing phenomena as they appear and function in the real world.

five themes of geography location, place, human-environmental interaction, movement, and region

flood inundation of an area not normally covered with water, caused by the temporary rise in water level in a river or other body of water.

floodplain a flat expanse of land that borders a river and becomes covered with water when the river overflows.

fresh water water that has a low salt content and is, therefore, drinkable, as distinguished from salt water in oceans and internal drainage systems; generally, water in lakes, rivers, and streams.

functional region a region that has some particular focus of spatial interaction within it (such as a school for a school district).

genealogy (jee nee AHL uh jee) an account of a person's descent from an ancestor or ancestors; a study that traces family descent. See family tree.

generalization a statement that expresses characteristics that exist in most, but not necessarily all, cases; particularly to describe spatial patterns or relationships of phenomena.

geographic grid system of intersecting lines (parallels of latitude and meridians of longitude) used to determine absolute location. See grid.

geographic past the spatial study of a particular location or area as it appeared at some time in the past.

glacier (GLAY sher) a large body of perennial ice that moves slowly down a slope.

gradient (GRAYD ee uhnt) the slope or rate at which the elevation of a river or stream decreases.

graphic scale a bar scale on a map which can be used to measure distances.

grid See geographic grid.

groundwater water that collects beneath the earth's surface, generally within 4 km (2.5 miles) of the surface and is replenished by surface water seeping into the earth.

hemisphere (HEHM uh sfier) one half of the earth, usually divided into northern and southern hemispheres and eastern and western hemispheres.

human activity interactions of humans with the physical environment.

human-environmental interaction adaptations to or changes in the natural environment made by humans.

human place characteristics the elements of the landscape that have been placed there by humans; human modifications of physical features.

hydroelectricity (hi droh ih lehk TRIHS uht ee) the conversion of the energy of moving water into electricity.

hydrologic (hi druh LAHJ ihk) **cycle** the exchange of water among oceans, the lower atmosphere, the land surface, and reserves beneath the earth's surface.

ice cap large, permanent mass of ice that generally extends over land surfaces, but may project over adjacent water areas (as in Antarctica).

immigrant (IHM ih gruhnt) a person who comes to a new country or area to permanently settle.

immigration (ihm uh GRAY shuhn) the process by which people come into a new country or area.

index contour heavy brown contour line that appears every fifth line on a topographic map.

inland sea a body of water characterized by internal drainage.

intermediate directions NNE, NE, ENE, ESE, SE, SSE, SSW, SW, WSW, WNW, NW, and NNW. (Compare with cardinal directions.)

internal drainage a drainage system not connected to oceans by rivers or streams.

interrelationship the mutual influence of one phenomenon on another.

island a land area that is smaller than a continent and surrounded by water.

journal a daily account of happenings; a record of personal observations and thoughts about particular topics.

lake an extensive body of water surrounded by land; a body of water smaller than oceans and most seas.

landscape the total of human and cultural features of a region.

latitude the distance in degrees north or south of the equator.

levee See artificial levee and natural levee.

line symbol a line on a map that represents a physical object (such as a highway or power line) or an idea, such as latitude and longitude or political boundaries.

local relief the difference in elevation between the highest and lowest points at a particular location or within a particular area.

location relative or absolute position on the earth's surface.

lock a chamber within a dam structure in which the water level can be adjusted to accommodate the passage of boats.

longitude (LAHN juh tyood) the distance in degrees east or west of the prime meridian.

map a symbolic representation of the world or a part of it.

map scale the ratio of distance on a map to the actual distance that is represented.

map symbol an abstract representation on a map of some cultural or natural feature. Types of symbols include point, line, area, and sometimes volume.

marsh poorly drained area temporarily covered with water.

material culture tangible objects made by humans; the variety of forms of such objects is a reflection of non-material culture (ideas).

meandering (mee AN duh ring) **river** a river with a curved channel that winds laterally across a floodplain.

melting pot a society that blends the characteristics of several diverse cultures or ethnic groups.

meridians (muh RID ee uhns) lines running from pole to pole on a map or globe, usually referred to as meridians of longitude.

migration the movement of a group of individuals in a species, including humans, from one location or region to another.

minute the 60th part of a degree of latitude or longitude.

monsoon a wind characterized by a seasonal reversal of direction. The summer heating of the continental mass causes a low pressure system and creates the wet monsoon (wind from water to land); the land-based high pressure of winter causes the dry monsoon (wind from land to water); monsoons are especially pronounced in the tropics, particularly in Asia.

mosaic a society composed of many cultural or ethnic groups who retain their differences and may occupy distinct areas.

mouth the point at which a stream or river discharges into a larger body of water.

movement the transfer of people, ideas, or goods from place to place.

natural levee a low ridge along a stream or river, formed by sediments deposited during flooding.

NIMBY acronym for "not in my backyard," an attitude of people who do not want development of some human activity in the vicinity of their homes.

nonmaterial culture the ideas, beliefs, customs, and traditions that make one group distinct from another.

oxbow lake a crescent-shaped lake beside a meandering river formed when the river cuts a new channel.

parallels the imaginary east-west lines around the earth, parallel with the equator, which show locations north and south of the equator.

pattern spatial distribution of geographic facts.

personality of place the combination of cultural and natural features of a particular location that create a unique character.

perspective a point of view that provides a frame of reference from which to ask questions; acquire, organize and analyze information; and answer questions.

phenomenon (fih NAHM uh nahn), plural, phenomena (fih NAHM uh nuh) an observable fact or event.

physical place characteristics the elements of the landscape, such as landforms, climate, vegetation, and soils, that result from natural processes.

place natural and human characteristics of a location.

pluralism See cultural pluralism.

point bar accumulated sediment on the inside of a river meander.

point symbol a map symbol used to designate the location of an object at a particular location or point, such as small black square that represents a house or other building on a map.

pollution undesirable change in the physical, chemical or biological characteristics of air, water, or land that can be harmful to living organisms.

pond a body of standing water, smaller than a lake and sometimes artificially formed.

population distribution variation of population density over a given country, region, or other area; where people live and in what numbers.

precipitation (prih sihp uh TAY shuhn) the deposits of water that reach the earth's surface from the atmosphere.

prime meridian the meridian of longitude assigned the value of 0 degrees longitude, which runs north and south through Greenwich, England, and from which locations are measured east or west.

pull factors desirable conditions that draw migrants from one location to another.

push factors undesirable conditions from which people escape by moving away.

quadrangle the area of earth's surface represented by a U.S. Geological Survey topographic map, the most common of which are 7.5-minute series (1:24,000) or 15-minute series (1:62,500).

rapids that portion of a river characterized by faster than normal flow, which is caused by a sharp gradient change, resistant rock in the channel, or both.

recycling the collection and reuse of waste materials.

region an area on earth's surface differentiated from surrounding areas by one or more specific characteristics.

regional study a geographic study of some area that possesses some degree of identity. Such a study is synthetic rather than analytic (as a systematic or topical study would be).

relative location a position on earth that is indicated with reference to the locations of other phenomena.

relief the configuration of the earth's surface.

representative fraction the scale of a map proportionate to the area represented on the earth's surface, shown as a ratio (for example, 1:24,000).

reservoir a body of stored water; humans create artificial reservoirs by building dams to regulate flow and produce hydroelectricity.

riprap large stones or pieces of concrete placed on the sloping banks of a river to prevent erosion.

river a body of water that flows into a larger body of water.

river system a river and its tributaries that carry the water of a drainage basin.

role playing (as used in Lesson 5) assuming the identity and characteristics of another person or a hypothetical person in order to better understand a point of view other than one's own. As part of a simulation, role players interact to explore social issues.

runoff the lateral movement of water to streams.

saline lake a body of water, such as the Great Salt Lake in Utah, that has internal drainage or is not connected to the ocean. The salinity or saltiness increases as water evaporates.

salinization (sal uh nuh ZAY shuhn) the excessive accumulation of salts in arid and semiarid regions caused by high evapotranspiration. Irrigation compounds salinization, and high levels of salt accumulation may make agriculture impossible.

sanitary landfill a specially prepared site for the deposition of solid waste. Beneath the landfill is an impermeable barrier to prevent leaching of harmful substances into the soil and groundwater. The top is covered with a layer of soil.

scale See map scale.

second the 60th part of a minute.

sediment eroded materials transported by the river.

sequent occupance (SEE kwuhnt AHK yuh puhns) a geographic study of a location or region at several historic times to assess changes from one period to another.

settlement facilities constructed by humans while occupying a region or place; a community of relatively recent development.

simulation an artificially created condition, designed to imitate reality for the purposes of training and exploring more directly the interactions of people, natural processes, or both.

soil moisture water in the soil layer that is available to plants.

source the beginning of a river or stream.

spatial (SPAY shuhl) **distribution** patterns resulting from the way geographic phenomena are distributed across part or all of earth's surface.

spatial interaction interdependence of phenomena (both physical and human); the patterns that show the physical relationships among phenomena; commonly used to refer to the movement of people, goods, and ideas from place to place.

special-interest group persons bound together to express their collective preference for particular interests. Such groups may be organized to promote legislation favoring their interests.

spring a continuous natural flow of water from the ground, occurring at the point where the water table intersects the land surface.

stated scale an expression of the ratio of distances on a map to distances on the area the map illustrates (for example, one inch = 12 miles).

stream a general term describing all scales of flowing water, though often used to refer to relatively a small body of flowing water.

subculture a distinct culture group that exists within a larger culture area.

swamp permanently waterlogged area characterized by trees or woody shrubs.

symbol See map symbol.

synthesis the process of combining elements to arrive at a composite picture, as in a regional study.

systematic or topical study a geographic study that focuses on individual subject matter components; it is analytic rather than synthetic.

temperature the amount of heat contained in matter, such as air or water; expressed in degrees on the Celsius (C) or Fahrenheit (F) scale.

topographic (tahp uh GRAF ihk) **map** a map with a sufficiently large scale to show detailed natural and cultural surface features.

topography (tuh PAHG ruh fee) the human and physical surface features of an area.

transpiration the process by which moisture in plants is changed to vapor.

tributary (TRIB yuh tehr ee) a stream or river that contributes its water to a larger stream or river.

uniform region a region in which the defining topic applies with significant similarity throughout.

valley the long, narrow depression created by the erosive force of a river or stream.

vegetation cover the amount and kinds of plant life found in a region.

velocity the rate at which water flows.

waste the discarded materials of living creatures, particularly humans, which often contribute to pollution.

waterfall the steep fall of a river, which occurs at an abrupt change in river gradient.

watershed the area drained by a river system; drainage basin.

wetlands temporarily saturated areas, marginal to dry land and open water, such as marshes and bogs.

wet monsoon moisture-bearing wind bringing heavy rainfall to an area.

wind the movement of air, caused by the uneven heating and cooling of earth's surface.

wing dam an artificial low-lying structure extending perpendicularly from a river bank to maintain a navigation channel.